Autonomous Mobile Mapping Robots

Edited by Janusz Będkowski

Published in London, United Kingdom

Autonomous Mobile Mapping Robots
http://dx.doi.org/10.5772/intechopen.100670
Edited by Janusz Będkowski

Contributors
Fabian Arzberger, Andreas Nüchter, Jasper Zevering, Anton Bredenbeck, Dorit Borrmann, Helge Andreas Lauterbach, SeungHwan Lee, Piotr Skrzypczyński, Kirill Krinkin, Anton Filatov, Elzbieta Roszkowska, Michal Drwiega, Janusz Będkowski, Jacek Szklarski

First published in London, United Kingdom, 2023 by IntechOpen
IntechOpen is the global imprint of INTECHOPEN LIMITED, registered in England and Wales, registration number: 11086078, 5 Princes Gate Court, London, SW7 2QJ, United Kingdom

British Library Cataloguing-in-Publication Data
A catalogue record for this book is available from the British Library

Additional hard and PDF copies can be obtained from orders@intechopen.com

Autonomous Mobile Mapping Robots
Edited by Janusz Będkowski
p. cm.
Print ISBN 978-1-83768-008-5
Online ISBN 978-1-83768-009-2
eBook (PDF) ISBN 978-1-83768-010-8

Meet the editor

Janusz Będkowski is a distinguished scholar with a DSc and Ph.D. in mobile robotics. Currently, he is a researcher at the Institute of Fundamental Technological Research, Polish Academy of Science. Dr. Będkowski has a wealth of experience in the practical and theoretical aspects of simultaneous localization and mapping applications on a global scale. His current work focuses on the use of autonomous mobile mapping methodologies to integrate geodesy, cartography, geoscience, and mobile robotics from an end-user perspective. Dr. Będkowski actively participates in the European Land Robotic Trial (ELROB) and the European Robotics Hackathon (ENRICH). Additionally, he contributes to open-source projects such as: https://github.com/JanuszBedkowski/gpu_computing_in_robotics, https://github.com/JanuszBedkowski/observation_equations, and https://github.com/MapsHD.

Contents

Preface

Autonomous mobile mapping robots produce digitized maps of their environment and are typically equipped with lidar, a camera, an inertial measurement unit, odometry, and sometimes GNSS for precise positioning. One of the challenges for these machines is simultaneous localization and mapping (SLAM), which is regarded as a "chicken and egg" problem: on one hand, the robot needs to be aware of its trajectory in order to construct a map; on the other hand, it needs to know the map in order to perform localization. There are several types of robots: unmanned ground vehicles (UGV), unmanned aerial vehicles (UAV), and unmanned surface vehicles (USV). Recent research has tended to equip all these robots with the same sensory setups, enabling common SLAM technology to be applied in all domains: air, ground, underground, and surface.

Air and surface applications can use GNSS and GPS efficiently, as there are usually no obstacles. The biggest challenge for autonomous mobile mapping ground robots is to combine all functionalities in a common framework, enabling missions to be executed without collisions with obstacles. Another important challenge is the limited mobility of wheeled robots. Recent work on ledge-climbing robots shows that these robots are capable of performing complex autonomous mapping missions, such as underground mining mapping, nuclear plant mapping, and many others. Thus, ledge-climbing robots are desirable for ground and underground autonomous mobile mapping tasks. In the air, the problem of obstacles affecting the mission is not present. Recent commercial drones are capable of collecting data autonomously or even performing SLAM on board. Challenges remain, such as underground mapping using drones. Fortunately, recent advances in on-board on-line lidar odometry provide efficient solutions for localization in hazardous environments.

This book consists of eight chapters in four sections. Following an introduction, the first section investigates alternative approaches to mobile 3D scanning. The second section considers key software components used for building autonomous mobile robots, including lidar odometry, loop closure, pose graph SLAM, map refinement, path planning, and coverage algorithms. The third section discusses multi-robot mapping approaches and scalable algorithms used in SLAM. The fourth section describes important urban search and rescue applications of aerial 3D mapping and automotive SLAM in urban environments.

I would like to thank Prof. Andreas Nuechter and Prof. Piotr Skczypczyński for their detailed contributions on real-world experiments.

Janusz Będkowski
Polish Academy of Science,
Institute of Fundamental Technological Research IPPT,
Warsaw, Poland

Section 1

Introduction and Future Challenges

Chapter 1

Introductory Chapter: Autonomous Mobile Mapping Robots – Current State and Future Real-World Challenges

Janusz Będkowski

1. Introduction

The concept of autonomous mobile mapping robots is composed of many technological advances, such as AI (Artificial Intelligence), sensors, locomotion, path planning, and data fusion. A great example of the research journey to autonomous robots is the autonomous car concept related to a certain level of autonomy. The first level assumes the driver support functionalities improving overall safety and the last level of autonomy assumes no driving wheel inside the vehicle. Obviously, mapping the world by single car is rather impossible due to the coverage required to build such digital representation. For this reason, autonomous cars will use existing maps and, if necessary, they will provide updates to them. A great example of an autonomous mobile mapping robot is a drone capable of executing flying missions and delivering data for further offline 3D map processing. This robot flies autonomously truth a predefined path and records all necessary data. Most of these robots are not capable of avoiding obstacles, but it is not relevant since the sky is rather empty space in most of the applications. Another example of an autonomous mobile machine is a cleaning robot performing cleaning mission on a predefined path. In this application, robot could be equipped with AI (Artificial Intelligence) capability to redefine the path according to sudden events, such as obstacle, not defined dirt on the path, etc. Most of the potential applications of the autonomous mobile mapping robots are related with the scenarios where human activities have to be reduced or even impossible to perform, such as space exploration or NPP (Nuclear Power Plant) inspection, during high radiation determined by accident.

Recent advances in robotic perception, such as non-repetitive scan pattern lidars [1], show a great decrease in the lidar cost, making it an affordable solution for massive autonomous robotic mobile mapping applications, such as aerial 3D mapping [2]. Non-repetitive scan pattern lidars are comparable with multi-beam lidars within the context of accuracy and temporal stability [3]. It is evident in the literature a great interest in this affordable perception plays an important role in expanding building 3D maps [4] in many applications. During last years, it was not so obvious that these narrow field-of-view lidars will be competitive with multi-beam lidars. Due to an integrated IMU (Inertial Measurement Unit), it is possible to perform online lidar odometry [5] without any additional engineering effort. Thus, recent advances

IntechOpen

in mobile mapping algorithms show great improvement in lidar 3D mapping even looking from the perspective of other applications such ADAS (Advanced Driver Assistance Systems) [6] (related to future autonomous driving).

Autonomous mobile mapping robots play important roles in many commercial applications, such as aerial mapping, underground mapping, search and rescue applications, space exploration, delivery robotics, and many others. These applications produce 3D maps with the trajectory as a core component. A trajectory is a set of consecutive poses (translation and rotation) with timestamps. These timestamps are crucial for retrieving poses for every measurement. Once we have the trajectory and calibration parameters of the mounted sensors (lidars, IMU, and cameras), it is possible to reconstruct the map. It is important to mention that we do not consider that the single source of truth (e.g., derived from GNSS signal) exists, thus a SLAM (Simultaneous Localization and Mapping) problem-solving is considered. Obviously, on one hand, we can consider GNSS as ground truth, on the other hand, we can find in the literature a comprehensive study that such a single source of truth can be improved by a data fusion approach [7]. Autonomous mobile robots equipped with multi-modal sensors (lidar, camera, and IMU) should consider the fact that a single source of truth does not exist, thus these machines should be equipped with the capability of data fusion. This solution provides an optimal result by combining trajectory calibration, observations, and maps.

2. Historical perspective

The core component of the autonomous mobile mapping robot is SLAM capability [8, 9]. The term SLAM [4, 5] corresponds to the so-called "chicken and egg dilemma"—what was first the chicken or the egg? For this reason, at the same time, robot should be equipped with efficient map representation to localize within this map, and at the same time the robot should provide accurate localization for building the map. SLAM is considered as an already solved mathematical problem, but it is evident looking at many robotic challenges that an efficient implementation does not exist. Recent advances in lidar technology show a great positive impact.

3. Real-world challenge

Mapping of large constriction sites and large urban buildings is a potential application for autonomous mobile mapping robots. An example is the boiling water reactor of the Zwentendorf Nuclear Power Plant (NPP Zwentendorf, **Figure 1**), which is the world's only nuclear power plant that has been completed but never put into operation. Thanks to EnRiCH [10] robotic trial it is evident that Zwentendorf areas are easily accessible, which in other NPPs can only be visited under severe difficulties. In active nuclear power plants extensive safety precautions are needed for human personnel due to the high level of radioactivity. Instead, in the NPP Zwentendorf engineers have transformed the plant and turbine halls into a training facility. Repair and dismantling measures but also critical incidents and disaster scenarios can be trained under realistic conditions, also autonomous mobile mapping robots are tested once every 2 years in so-called EnRiCH—European Robotics Hackathon [10].

To demonstrate the complexity of the mapping challenge **Figure 2** shows photos from a mapping survey performed with an affordable mobile mapping backpack system [11]. **Figure 3** shows that more than three-kilometer length trajectory is

Figure 1.
Large urban building—the boiling water reactor of the Zwentendorf Nuclear Power Plant.

Figure 2.
Mapping survey performed with an affordable mobile mapping backpack system [11].

required to cover the entire scene. The result of the mapping—registered 3D point cloud representing the Zwentendorf Nuclear Power Plant is shown in **Figures 4** and 5.

This challenge shows fundamental requirements for an autonomous mobile mapping robot that should be capable traverse a 3.5 km length trajectory, including 15 levels of stair. At the current stage of robotic technology, only a swarm of ground and air robots can reach such a level of autonomy and coverage.

4. Conclusion

On one hand, we can observe that SLAM problem is already solved, on other hand many real-world challenges [12, 13] in autonomous mobile mapping robots prove

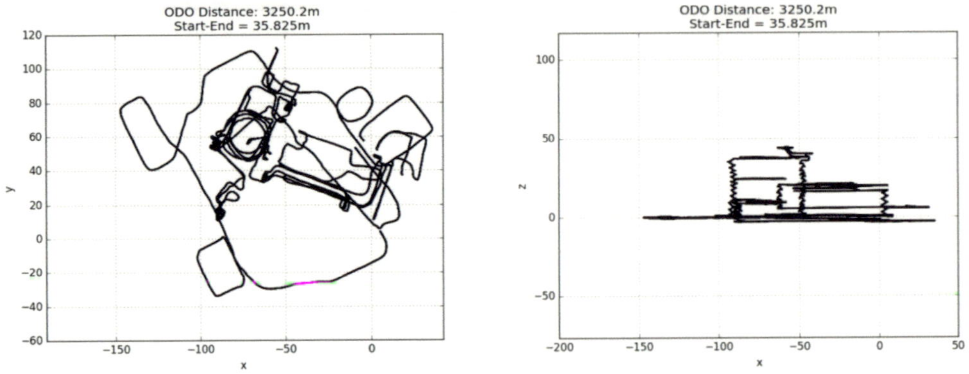

Figure 3.
The more than three-kilometer length trajectory required to cover the entire scene of the Zwentendorf Nuclear Power Plant.

Figure 4.
Perspective and top view, the result of the mapping—registered 3D point cloud representing the Zwentendorf Nuclear Power Plant.

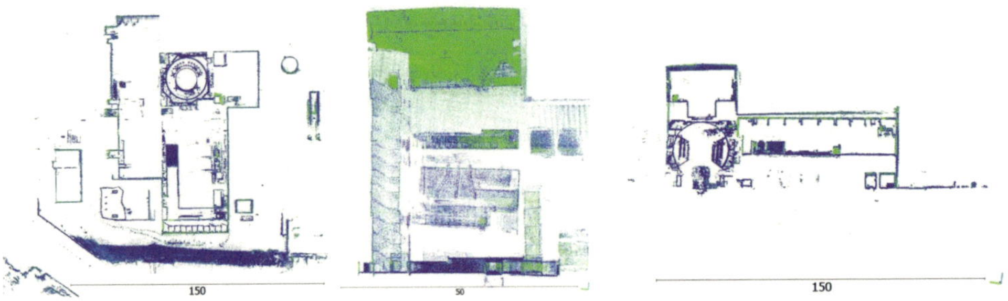

Figure 5.
Result of the mapping—registered 3D point cloud representing the Zwentendorf Nuclear Power Plant.

great interest in this domain. The main problem is the cost of the robot capable of performing missions in hazardous environments. For this reason, this research activity is not so popular, therefore the technological improvements are rather incremental than revolutionary. An opportunity is in the autonomous car driving domain and delivery robotics since it requires collecting high volume, large scope data, and executing SLAM by many agents. To conclude, autonomous mobile mapping robots are fascinating and require many efforts to deliver satisfactory results.

Author details

Janusz Będkowski
Polish Academy of Science, Institute of Fundamental Technological Research IPPT, Warsaw, Poland

*Address all correspondence to: januszbedkowski@gmail.com

References

[1] Jiahe C, Jianwei N, Zhenchao O, Yunxiang H, Dian L. ACSC: Automatic calibration for non-repetitive scanning solid-state LiDAR and camera systems. arXiv. 2020

[2] Aldao E, González-de Santos LM, González-Jorge H. LiDAR based detect and avoid system for UAV navigation in UAM corridors. Drones. 2022;**6**:185. DOI: 10.3390/drones6080185

[3] Kelly C, Wilkinson B, Abd-Elrahman A, Cordero O, Lassiter HA. Accuracy assessment of low-cost lidar scanners: An analysis of the Velodyne HDL–32E and Livox Mid–40's temporal stability. Remote Sensing. 2022;**14**:4220. DOI: 10.3390/rs14174220

[4] Wang Y, Lou Y, Zhang Y, Song W, Huang F, Tu Z. A robust framework for simultaneous localization and mapping with multiple non-repetitive scanning lidars. Remote Sensing. 2015;**2021**:13. DOI: 10.3390/rs13102015

[5] Tixiao S, Brendan E, Drew M, Wei W, Carlo R, Daniela R. LIO-SAM: Tightly-coupled lidar inertial odometry via smoothing and mapping. In: IEEE/RSJ International Conference on Intelligent Robots and Systems (IROS). Las Vegas, NV, USA (Virtual). 2020. pp. 5135-5142

[6] Long N, Yan H, Wang L, Li H, Yang Q. Unifying obstacle detection, recognition, and fusion based on the polarization color stereo camera and LiDAR for the ADAS. Sensors. 2022;**22**:2453. DOI: 10.3390/s22072453

[7] Bedkowski J, Nowak H, Kubiak B, Studzinski W, Janeczek M, Karas S, et al. A novel approach to global positioning system accuracy assessment, verified on LiDAR alignment of one million kilometers at a continent scale, as a foundation for autonomous driving safety analysis. Sensors. 2021;**21**:5691. DOI: 10.3390/s21175691

[8] Leonard J, Durrant-Whyte H. Simultaneous map building and localization for an autonomous mobile robot. In: Proceedings IROS'91:IEEE/RSJ International Workshop on Intelligent Robots and Systems'91. Vol. 3. Osaka, Japan: IEEE; 1991. pp. 1442-1447. Date of Conference: 03-05 November 1991, Date Added to IEEE Xplore: 06 August 2002. Print ISBN: 0-7803-0067-X. INSPEC Accession Number: 4199151. DOI: 10.1109/IROS.1991.174417.

[9] Skrzypczynski P. Simultaneous localization and mapping: a feature-based probabilistic approach. International Journal of Applied Mathematics and Computer Science. 2009;**19**(4):575-588

[10] Available from: https://enrich.european-robotics.eu

[11] Pelka M, Majek K, Bedkowski J. Testing the affordable system for digitizing USAR scenes. In: 2019 IEEE International Symposium on Safety, Security, and Rescue Robotics (SSRR). Germany: Julius-Maximilians University of Würzburg, IEEE Press; 2019. pp. 1-2. DOI: 10.1109/SSRR.2019.8848929

[12] Available from: https://research.csiro.au/robotics/our-work/darpa-subt-challenge-2018/

[13] Available from: https://hilti-challenge.com

Chapter 2

Unconventional Trajectories for Mobile 3D Scanning and Mapping

Fabian Arzberger, Jasper Zevering, Anton Bredenbeck, Dorit Borrmann and Andreas Nüchter

Abstract

State-of-the-art LiDAR-based 3D scanning and mapping systems focus on scenarios where good sensing coverage is ensured, such as drones, wheeled robots, cars, or backpack-mounted systems. However, in some scenarios more unconventional sensor trajectories come naturally, e.g., rolling, descending, or oscillating back and forth, but the literature on these is relatively sparse. As a result, most implementations developed in the past are not able to solve the SLAM problem in such conditions. In this chapter, we propose a robust offline-batch SLAM system that is able to address more challenging trajectories, which are characterized by weak angles of incidence and limited FOV while scanning. The proposed SLAM system is an upgraded version of our previous work and takes as input the raw points and prior pose estimates, yet the latter are subject to large amounts of drift. Our approach is a two-staged algorithm where in the first stage coarse alignment is fast achieved by matching planar polygons. In the second stage, we utilize a graph-based SLAM algorithm for further refinement. We evaluate the mapping accuracy of the algorithm on our own recorded datasets using high-resolution ground truth maps, which are available from a TLS.

Keywords: 3D LiDAR, mobile mapping, scanning, spherical robot, pendulum, descent, small FOV, Livox, Intel, RealSense

1. Introduction

Mobile systems increasingly gain astonishing capabilities when it comes to 3D sensing, mapping, and environment reconstruction (**Figure 1**). Nowadays, there exist many mobile systems in different shapes and sizes that are able to perform these tasks, e.g., drones, wheeled or tracked robots, or backpack-mounted systems, just to name a few. A key aspect of fulfilling their purpose is the estimation of the systems inertial frame of reference, i.e., the systems local coordinate system, with respect to a global reference frame. This global reference frame has an origin somewhere in the respective environment and is usually initialized with the systems starting position and orientation (pose). By expressing the local measurements in the global frame, the system is able to create a map of the environment whilst localizing itself in it. This process is called simultaneous localization and mapping (SLAM), and only works if the initial pose estimation of the system is sufficiently accurate. The types of robot

Figure 1.
(Left) post-processed 3D point cloud acquired by a sensor that is rotating freely while being descended with a crane from a fire truck. Prior pose estimates are available from three IMUs and an angular encoder, which encodes the rotation of the cable reel. (Center) the firetruck. (Right) the sensor, which is mounted to the crane of the firetruck.

designs mentioned earlier are often favored since they have access to quite accurate prior pose estimates (GNSS, odometry, visual odometry, etc.) and reliable coverage while sensing the environment with cameras or laser scanners. In such conditions, it is typically easy to perform laser-based SLAM, either online (for autonomous mobile robots) or as a post-processing step. Yet, there are still many situations where conditions are poor, inferring uncertainties to prior pose estimates and thus degrading SLAM performance. Visual feature tracking with a camera, for example, only works in good lighting situations, and GNSS might not be available. Using IMUs as the only pose estimation device usually is also not an option, due to the large accumulation of measurement errors caused by noise and drift which makes position estimation difficult. Even high-end devices, e.g. in aviation systems, must be combined with other references like GNSS in order to be reliable. The ability of LiDAR-based SLAM algorithms to deal with degradation puts constraints on current mobile system designs, as well as their applications. Therefore, extending the capabilities of SLAM algorithms to more unusual scenarios opens up interesting design choices for mobile systems, especially in hazardous environments. The intention of this chapter is to provide a general and robust LiDAR-based solution for the SLAM problem, which is independent with respect to the executed trajectories and sensor setups. We note the utilization of a Livox Mid-40 scanner, which is considered a solid-state LiDAR in [1]. Yet in [2], the family of Livox scanners is only considered to be "semi solid-state" LiDARs, due to their non-repetitive scanning pattern. Solid-state LiDAR got a lot of attention in the past years due to "their superiority in cost, reliability, and [...] performance against the conventional mechanical spinning LiDARs [...]" [3]. While traditional LiDAR is based on electro-mechanic parts which move the sensor head, solid-state LiDAR relies on micro-electro-mechanical systems (MEMS), optical-phase arrays (OPA), or Risley prisms. Despite their potential advantages, solid-state laser-scanners impose new challenges for established SLAM algorithms, e.g., small FOV, an irregular scanning pattern with non-repetitive scanning which makes feature extraction more difficult, and increased motion blur. In this work, we address these challenges by making the assumption that planar structures are available in the environment. Due to the utilization of planes, the envisioned applications are in human-made environments, e.g., old buildings that are in danger of collapsing, narrow underground tunnels, construction sites, or mining shafts. Many mobile mapping systems present in the current literature have access to accurate pose estimates, as well as good sensing coverage. We present four notably interesting experimental setups which do not meet those conditions. The main sensor in each scenario is a LIVOX MID-100 laser scanner, which is inconvenient due to its limited FOV and unique scanning pattern. **Figure 2** illustrates the different executed motions. (1) We descent a freely rotating 3D sensor from a crane. Prior pose estimates are available from three IMUs and a rotary encoder on the cable reel. (2) Further, we use the sensor as a pendulum, oscillating back and forth while walking. The robot obtains prior pose estimates from a T265 tracking camera, which uses its own internal IMU. (3) Moreover, we roll the sensor around on flat ground. An IMU-only-based approach with three units estimates the pose of the system [4]. The approach combines two popular filters (Madgwick- and Complementary-filter) and estimates the position by relating the rotational velocity and radius of the sphere to its traveled distance. Further constraints are present in the filter to account for slippage and sliding effects of the sphere. (4) Finally, we repeat the previous experiment but substitute the 3D sensor with a 2D sensor, which measures parts of the environment in planar slices. Rotating the planar slice thus results in

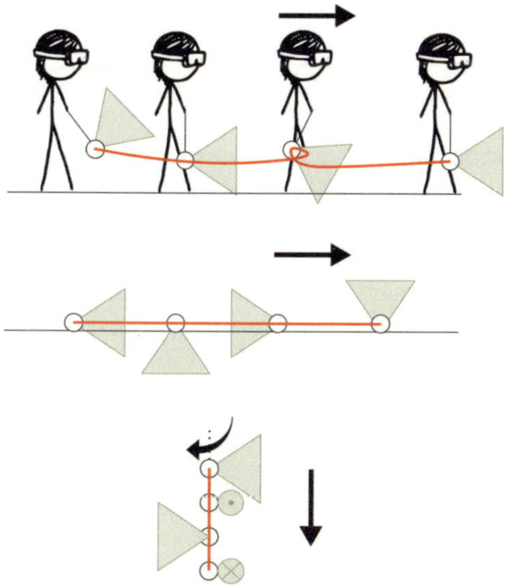

(a) Different motion profiles for scanning. The red line shows the position of the robot over time. The green cones represent the scanning direction and angle of the sensor. (Upper) Walking while swinging the pendulum setup. When the pendulum swings back while moving forward it creates the "looping" shape of the trajectory. (Center) Rolling on flat ground. (Lower) Descending while rotating. The green circles symbolize that either the scanner is facing towards us or away from us.

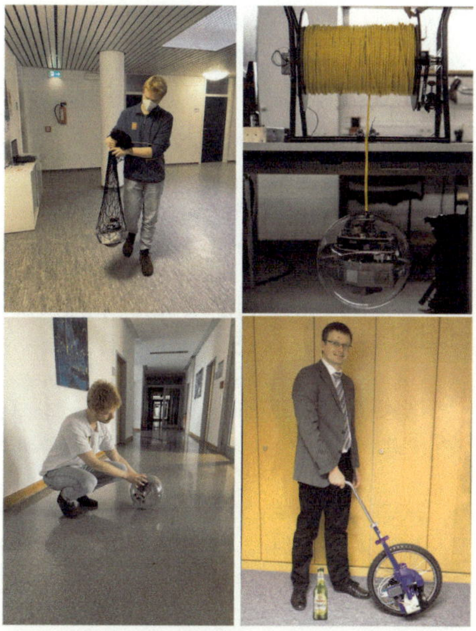

(b) Setup for the experiments. Upper left: Pendulum setup in the hallways of an office environment. We put the sensors inside a trailer net and swing it back and forth while walking. Upper right: Descending setup with a tear-resistent tether cable, which is rolled around a coil. Bottom left: Rolling on flat surfaces setup in the hallways of an office environment. The sensors are placed inside a protective spherical plastic shell. Bottom right: Rolling on flat surfaces setup with the RADLER [4] device.

Figure 2.
Illustration of the executed sensor trajectories (left) and images of system setups during operation (right).

a 3D reconstruction of the scene. For pose estimates, we use a rotary encoder and a single IMU. Section 4 describes all experimental setups in more detail. The prior pose estimates are subject to a significant amount of noise and drift. Some scenarios are more difficult than others since the robot locomotion mechanism causes the sensor to move in five, or even all six degrees of freedom (DOF). In this work, we refine our offline-batch SLAM system from [5] to make it more robust, such that it is able to address the unfavorable conditions imposed by more challenging trajectories. Section 3 introduces the two-staged SLAM algorithm, where the first stage uses a polygon-based approach for a fast coarse alignment, and a graph-based method in the second stage for slow further refinement. We evaluate the accuracy of the resulting maps by means of ground truth point clouds, which are available from high precise terrestrial laser scanning (TLS). Note that in this initial study, no trajectory ground truth has been recorded. This is a task for future work. In particular, the contributions of this work are as follows:

- Fixing the open problem mentioned in previous work [5], where after the application of the SLAM algorithm the resulting trajectories were jagged instead of smooth.
- Reducing the number of hyperparameters for our SLAM algorithm [5] by seven, without the sacrifice of flexibility or performance. We achieve this by substituting the iterative optimization using AdaDelta with a closed-form solution based on singular value decomposition (SVD).
- Introducing major technical improvements regarding our SLAM procedure from previous work [5], which increase the robustness of the algorithm. These include the substitution of the local planar clustering (LPC) with a different plane detection framework according to [6], as well as the implementation of a simple global plane model that builds up sequentially as new range measurements arrive.
- Introducing a globally consistent graph-based method, "semi-rigid SLAM," according to [7] as an additional post-processing step, to further decrease the amount of accumulated registration errors.
- Publishing the challenging datasets themselves, as we encourage readers to try their own SLAM algorithms on them. A more detailed description of the datasets is found in Section 4.

2. Related work

Many state of the art 3D scanning and mapping approaches for mobile systems are based on wheeled robots, drones, or backpack-mounted solutions. On the other hand, the literature regarding mobile mapping systems with more unconventional trajectories is relatively sparse. In the most recent past, our lab has addressed the subject more frequently with the "RADial LasER scanning device" (RADLER) [8], the "Laser-Mapping Unidirectional Navigation Actuator" (L.U.N.A.) [9], and another publication regarding rolling 3D sensors in human-made environments [5]. RADLER is a modified unicycle where a SICK LMS141 2D scanner is fixed to the wheel. As far as we know it was one of the first systems where the same rotation that is used for locomotion is also utilized to actuate the sensor. L.U.N.A. extends this idea by employing a self-actuated locomotion approach using internal flywheels. Another noticeable trend, despite being only a concept so far, is the European Space Agency's (ESA) interest in spherical robots capable of SLAM, called DAEDALUS [10]. In their Concurrent Design Facility (CDF) study, we developed a mission to autonomously explore underground caves and lava tubes on the moon with DAEDALUS [11], which emphasizes the potential of this type of system design for hazardous environments. More examples of scanning mobile systems with unconventional trajectories include Zebedee [12], a handheld 2D range scanner mounted on a spring that estimates its pose using an IMU, or VILMA [13], an IMU-less rolling system that uses only range measurements for localization. It later advanced into a commercial solution called "ZEB-Revo" [14]. Leica also provides a handheld sensor for mobile mapping purposes with their BLK2GO [15]. Alismail and Browning [16] provide a marker-less calibration procedure for spinning actuated laser scanners, where the extrinsic parameters with respect to the spinning axis are estimated. In this initial study, we go without fine calibration of extrinsic as the constant calibration errors are less significant than the errors introduced by the factors stated in the previous section. In terms of laser-based SLAM, many algorithms for 6 DOF are available, often based on the well-known Iterative-Closest-Point (ICP) algorithm [17]. Lu and Milos [18] derive a graph-based 2D variant that considers all scans simultaneously in a global fashion. Later, their approach got adopted for 3D point clouds in 6 DOF [19] and extended further to a semi-rigid continuous-time method [7]. This is the method we use in this work as an additional post-registration step, to reduce the amount of accumulated errors during the first scan-matching stage. Another recent continuous-time graph-based framework is "IN2LAAMA" [20], which is able to perform localization, mapping, and extrinsic calibration between a laser-scanner and IMU at the same time. It is an offline-batch method optimized for 360° FOV multichannel LiDAR devices and has been extensively tested with a Velodyne VLP-16, yet is not suited for the application to recently arising solid state laser-scanners. There also exist continuous-time graph-based *online* methods, such as [21] which organize and optimize the system poses using a multilevel hierarchical graph. This method achieves comparable accuracy as similar offline-batch methods by means of multiresolution surfel-based registration. However, the approach is also optimized for traditional multichannel laser scanners and has been tested on carefully controlled micro aerial vehicles (MAVs), which ensures good coverage. More approaches to laser-based SLAM exist that do not rely solely on point-to-point optimization as ICP does. Popular model-based optimization methods often deal with finding planes in the environment, as considering planes is more robust to outliers and noise than considering only points. In Ref. [22], Förster et al. successfully use plane-to-plane correspondences for optimization. Their approach assumes that planar patches got pre-extracted from the point cloud with a method of choice, and incorporates possible uncertainties in the plane model. Further recent examples of laser-based SLAM approaches making use of the existence of planes include [23–26]. Similar registration procedures to ours are [27], which uses plane-to-plane correspondences for pre-registration and point-to-plane correspondences afterward, and [28], which uses point-to-point, as well as plane-to-plane correspondences based on their availability. Two more recent and very interesting SLAM approaches which specialize more on the massivley produced LiVOX devices, are "Loam-livox" [3] and "Livox-mapping" [29]. The former is based on the well-known LOAM [30] algorithm, while the latter is provided directly by Livox. Both have been especially optimized for small FOV devices and offer a feature extraction approach which is suitable for the never repeating, flower-shaped scanning pattern. However, they have been designed with the intention of using them for autonomous driving cars, which follow simpler trajectories compared to this work.

3. SLAM approach

To address the unconventional trajectories, we use a flexible two-staged SLAM approach, which is described in this section. We initially proposed a version of the first stage in [5]. The approach is based on finding planar polygons in the scans and matching them against a global model. In this section, we build upon our previous work and introduce several changes. The second stage of our algorithm is "Semi-Rigid SLAM" [7], which further decreases accumulated registration error from the first stage. It is a continuous-time method where each frame is optimized simultaneously using a partially connected pose graph. **Figure 3** shows a block diagram representing the information flow of the SLAM system, including the additions made in this work. Some basic derivations stay the same (see [5] for further details). Let a point in 3D space be defined as $\mathbf{p}_i = (x_i, y_i, z_i)^\tau$. Further, a homogeneous transformation of that point along the translation $\mathbf{t} = (t_x, t_y, t_z)^\tau$ and rotation defined using the roll-pitch-yaw $(\varphi - \vartheta - \psi)$ Tait-Brian angles is given:

$$T(\mathbf{p}_i) = \begin{bmatrix} x_i C_\vartheta C_\psi - y_i C_\vartheta S_\psi + z_i S_\vartheta + t_x \\ x_i (C_\varphi S_\psi + C_\psi S_\varphi S_\vartheta) + y_i (C_\varphi C_\psi - S_\varphi S_\vartheta S_\psi) - z_i C_\vartheta S_\varphi + t_y \\ x_i (S_\varphi S_\psi - C_\varphi C_\psi S_\vartheta) + y_i (C_\psi S_\varphi + C_\varphi S_\vartheta S_\psi) + z_i C_\varphi C_\vartheta + t_z \end{bmatrix}, \quad (1)$$

where C_a and S_a denote cosine and sine with argument a. Additionally, let a plane in 3D space be defined as $\rho_k = \{\mathbf{n}_{\rho_k}, \mathbf{a}_{\rho_k}\}$, where $\mathbf{n}_{\rho_k}$ is the normal vector of the plane and $\mathbf{a}_{\rho_k}$ is its supporting point. The problem we must solve is an optimization problem, where the following error function E has to be minimized:

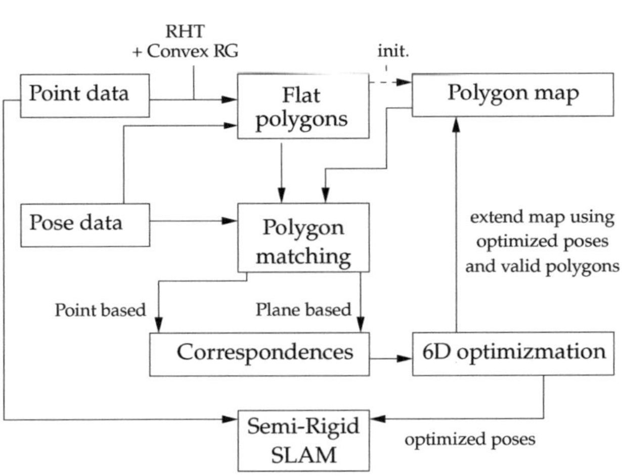

Figure 3.
Overview of the proposed SLAM system. The polygon map represents a set of flat, convex-shaped bounding boxes of dominant planes. This model is used to find similarities between polygons from subsequent sensor data.

$$E(T) = \sum_{\rho_k} \sum_{\mathbf{p}_i \in \rho_k} \| \mathbf{n}_{\rho_k} \cdot [T(\mathbf{p}_i) - \mathbf{a}_{\rho_k}] \|^2 \ , \tag{2}$$

Note that point-to-plane correspondences ($\mathbf{p}_i \in \rho_k$) have to be available, which we establish by matching polygons similar to [5]. However, the global polygon model in this work is not extracted only once as an initialization step, as in our previous work. It is instead a new global model, which is initialized and then updated sequentially. Our plane detection approach does also not rely on local planar clustering (LPC) anymore. The old method is a region-growing-based approach to cluster points with similar normals, whereas the new approach uses a Hough-transformation (HT) framework as in Ref. [6]. We describe the abovementioned additions in the following subsections in more detail.

3.1 Local and global plane model

As mentioned earlier, in our previous work [5], we rely on LPC to identify planes in each scan, as well as the points that belong to those planes. The approach calculates normal vectors for each point and clusters them into planar patches based on their distance and angle. Then, after each point in a scan was potentially identified to belong to one plane, correspondences have to be established with respect to the global model. In Ref. [5], we obtain the global model by extracting planes from only a few initial measurements according to Ref. [6].

In this work, we replace LPC with that same approach [6] to identify planes in each scan. The new approach is based on a randomized version of the well-known Hough transformation (see Algorithm 5 in Ref. [6]). After a plane has been identified in the Hough space, all the points belonging to that plane are considered, and their convex hull is calculated. However, instead of deleting all the points close to the newly identified plane, we save them in a point cluster and link it to the corresponding plane. That way, we are still able to establish point-to-plane correspondences as in our previous work. **Figure 4** illustrates how the extracted planes from each frame are used to sequentially update the global model. The upper sequence of **Figure 4a** shows the abovementioned point clusters with their corresponding planes for different frames. Note that in the last figure of the sequence, identical planes from the different frames are merged after registration. The bottom **Figure 4b** shows the resulting global plane model, as well as the point cloud after registration of all frames. Merging two planes works by considering all the points on both convex hulls, and recalculate the convex hull and normal vector from those points. Note that this is not fully a dynamic model, as polygons are added sequentially but never refined or even falsified after being added, leading to registration errors such as duplicate and misaligned walls. Now that we have a global plane model, as well as planar point clusters in each subsequent frame, we use the matching function from our previous work [5] to establish correspondences between the two. In the next subsection, we use these correspondences to minimize Eq. (2).

3.2 Closed-form solution with singular value decomposition

In our previous work [5], the minimization of Eq. (2) is achieved by the AdaDelta [31] method, which is based on analytical jacobians and stochastical gradient descent (SGD). However, SGD-based methods require multiple iterations until they converge

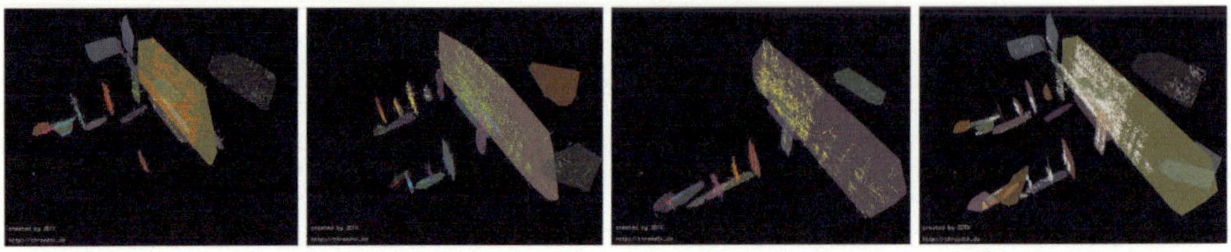

(a) From left to right: (1 - 3): Subsequent point cloud and pose data from a laser scanner. The points are grouped in planar clusters, represented by the point color, and a plane is fit through each cluster. (4) The global model that results from sequentially merging the plane observations from 1 - 3, using the first stage. Identical planes from different frames get merged after registration.

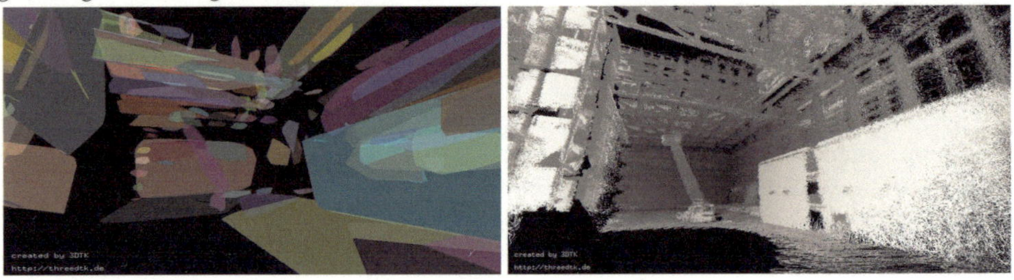

(b) (Left) Resulting global plane model, which gets created during the first stage of our SLAM approach. (Right) Corresponding point cloud from the same view for comparison. The grayscale values of the points represent reflectivity.

Figure 4.
Illustration of how the global plane model is obtained and sequentially extended from individual measurements.

to a solution. Furthermore, a hyperparameter is added for each of the six degrees of freedom, as well as an additional parameter specifying the maximum number of SGD iterations before updating correspondences. These parameters are not required anymore, as we now introduce a closed-form solution based on singular value decomposition (SVD). The first appearance of SVD in the context of point set registration is in Ref. [32]. The solution assumes that point-to-point correspondences exist, instead of point-to-plane correspondences. To this end, we must first calculate the projection point from our source point to the target plane. Therefore, the source point gets shifted onto the target plane in the direction of the plane's normal vector. Let D be the signed distance of the point to the plane in the normal direction:

$$D_k^i = \mathbf{n}_{\rho_k} \cdot \left[T(\mathbf{p}_i) - \mathbf{a}_{\rho_k}\right] \ . \tag{3}$$

Then, the projection point onto the plane is given as:

$$\mathbf{P}_k^i = T(\mathbf{p}_i) - D_k^i \cdot \mathbf{n}_{\rho_k} \ . \tag{4}$$

We use this point for point-to-point correspondence. However, we note that $\mathbf{P}_k^i$ is also on the corresponding plane, thus solving the point-to-point problem with SVD also minimizes our initial Eq. (2). First, we need the correspondence centroids, i.e., the centroid of all the plane projection points, and the centroid of all the data points. Let N be the number of correspondences, then the centroid of plane projections is as follows:

$$\mathbf{c}_m = \frac{1}{N} \sum_{\mathbf{p}_i \in \rho_k} \mathbf{P}_k^i \ , \tag{5}$$

and the centroid of data points is as follows:

$$\mathbf{c}_d = \frac{1}{N} \sum_{\mathbf{p}_i \in \rho_k} T(\mathbf{p}_i) \ . \tag{6}$$

We set up the real 3x3 correlation matrix **M** as follows:

$$\mathbf{M} = \sum_{\mathbf{p}_i \in \rho_k} \left(T(\mathbf{p}_i) - \mathbf{c}_d\right)\left(\mathbf{P}_k^i - \mathbf{c}_m\right)^{\tau} \ , \tag{7}$$

which is decomposed using SVD as follows:

$$\mathbf{M} = \mathbf{U}\Sigma\mathbf{V}^{\tau} \ . \tag{8}$$

The interested reader might consider [33] for a detailed description on how the decomposition works. We set up another real 3x3 rotation matrix **R**, which solves the rotation needed to minimize Eq. (2):

$$\mathbf{R} = \mathbf{V}\mathbf{U}^{\tau} \ . \tag{9}$$

From this, the translation that minimizes Eq. (2) is calculated as follows:

$$\mathbf{t} = \mathbf{c}_m - \mathbf{R}\mathbf{c}_d \ . \tag{10}$$

Note that this is a closed-form solution that does not need any hyperparameters. In contrast, AdaDelta [31] needs a hyperparameter for each of the 6 degrees of freedom plus an additional parameter for the maximum number of inner iterations.

3.3 Condense and atomize

Our previous work [5] mentions that after the application of the algorithm, i.e., the first stage in this work, the resulting paths look distorted. This is due to two reasons: (1) The individual frames are "condensed" into metascans, which are referenced with only one pose. We call a collection of multiple subsequent scans, which are represented using a shared coordinate system, a metascan. (2) Some scans do not contain any points due to minimum scanning range, e.g., when rolling over the floor, thus they are not optimized. We fix these problems by introducing the inverse operation to "condense," which is able to distribute the relative transformation that got applied to the metascans, back onto the individual scans. This back-distributing operation is what we call "atomize." **Figure 5** illustrates the process of condensing, registration in the condensed domain, and atomizing. During the condense operation, one has to transform all points from different frames in the same reference coordinate system. Note that we start counting the frames with one. Let J be the number of all frames, which should be condensed to a total number of M frames, where $M \leq J$. Let S be an arbitrary integer chosen from the interval $\left[1, \lfloor\frac{J}{M}\rfloor\right]$. The number S defines which frame we pick as a reference coordinate system for the points from the other scans in the interval. Next, consider all J frames, given has homogeneous 4x4 matrices,

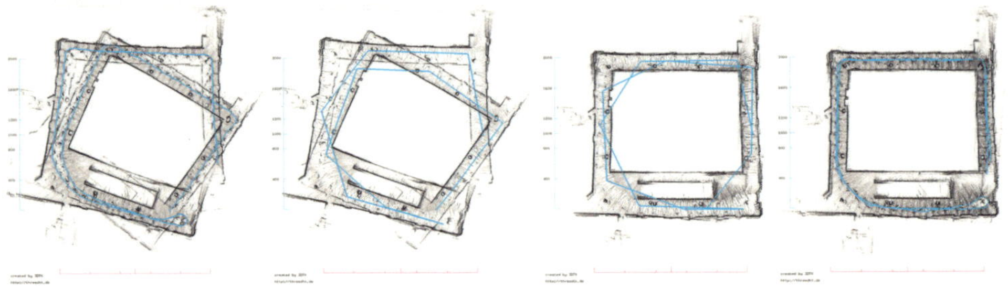

Figure 5.
Trajectories are drawn in blue and result from connecting every subsequent pose with a straight line. From left to right: (1) full point cloud with all initial poses. (2) "condensed" point cloud with only 12 poses. A total of 1000 linescans get combined into one metascan, which has its origin at the pose of the middlest scan. Further, we subsample and filter the metascans themselves. (3) the point cloud from 2 after registration. Only 12 poses got optimized in "condensed" domain. (4) full point cloud and all optimized poses after the "atomize" operation, which is the inverse to "condense." from the optimized poses in 3, we calculate the relative transformation that has to be applied to all initial poses.

$\{\mathbf{H}_1, \mathbf{H}_2, \cdots, \mathbf{H}_S, \cdots, \mathbf{H}_{2S} \cdots \mathbf{H}_J\}$. The frames with indices $m \cdot S$ (where $m \leq M$) are the indices of the metascan frames, where all points from the other frames, corresponding to the same interval, must be transformed in. Thus, the relative transformation between any single frame with index $j \leq J$ and metascan frame with index m is as follows:

$$\mathbf{H}_{m,j} = \mathbf{H}_{m \cdot S}^{-1} \mathbf{H}_j \quad . \tag{11}$$

We apply this transformation to every point in the j-th frame, for all J frames. Now we have a total of only M metascan frames $\mathbf{H}_{m \cdot S}$, i.e., $\{\mathbf{H}_S, \mathbf{H}_{2S}, \cdots, \mathbf{H}_{M \cdot S}\}$, which are input to our first SLAM stage. After the application of the first stage, there are M optimized frames $\hat{\mathbf{H}}_{m \cdot S}$, which we denote with a hat. The atomize operation has to apply the relative pose change between $\mathbf{H}_{m \cdot S}$ and $\hat{\mathbf{H}}_{m \cdot S}$, back onto the individual scans. Thus, the new optimized frames for all J original frames are as follows:

$$\hat{\mathbf{H}}_j = \left(\hat{\mathbf{H}}_{m \cdot S} \mathbf{H}_{m \cdot S}^{-1}\right) \mathbf{H}_j \quad . \tag{12}$$

After distributing the relative pose-change onto the individual scans in that way, the second stage of our approach begins. In the second stage, every individual frame is considered simultaneously in a pose graph.

3.4 Post-registration with semi-rigid SLAM

In our first SLAM stage, each scan is considered sequentially. That way, the algorithm creates a global plane model of the environment, without the knowledge of future measurements. This leads to registration inaccuracies during the first stage, which is why we use a second stage afterward: "Semi-Rigid Registration" (SRR) [7]. The method considers all scans simultaneously in a continuous-time fashion using a pose graph. In the graph, each pose is represented by a node and is connected via edges to other poses if the overlap between the corresponding scans is large enough. After one iteration of the algorithm, SRR re-calculates the edges. **Figure 6** illustrates the behavior of SRR on a dataset recorded by a spherical system rolling on a flat ground. Section 4.2 describes the experimental setup and evaluation of the dataset. In

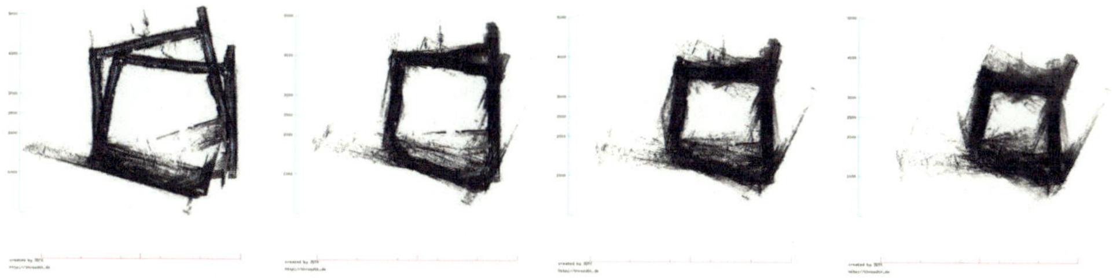

Figure 6.
Illustration of multiple iterations of "semi-rigid registration" (SRR) [7]. From left to right: (1) resulting point cloud with initial pose estimations, zero iterations. (2) resulting point cloud after 50 iterations of SRR. (3) after 100 iterations. (4) after 150 iterations.

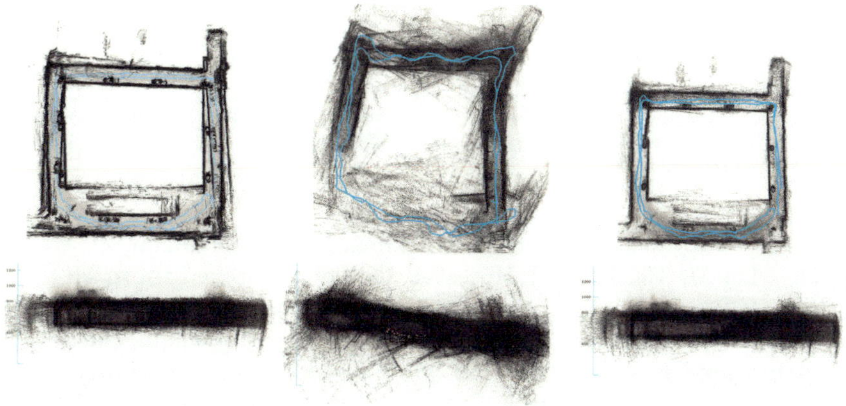

Figure 7.
Comparison of the first stage, second stage, and first stage followed by the second stage. Upper row: Point clouds in birds-eye view, with the sliced ceiling. Lower row: Point clouds in side view, aligned to the initial orientation. The initial input to the algorithms is shown in the most left image in ***Figure 6****. From left to right: (1) first stage only. (2) second stage (SRR) only. (3) first stage is followed by the second stage.*

the given example, the initial pose estimates are subject to a large amount of drift. Although SRR has been designed to also reduce such large-scale errors, the method alone is not able to perform well on more complex trajectories shown in this work. However, if the input to SRR is already coarsely aligned, the quality of point clouds and trajectories improves, as shown in **Figure 7**. The images in the left column show the resulting point cloud after the application of the first stage. The walls are not perfectly aligned, yet the side view reveals that the trajectory is planar. In the centered column, SRR is not able to register the measurements in a meaningful way, using the initial pose estimates. Moreover, the point cloud and trajectory are less planar. In the right column, both stages get applied, resulting in better overall map quality. Using the input from the first stage, the graph-based second stage is able to reduce the accumulated registration error from the first stage.

3.5 Comparison with ICP

The previous section has demonstrated that using SRR alone is not an option with the given trajectories. **Figure 6** shows how SRR is not able to converge to a meaningful solution when being applied to a dataset from a rolling spherical system. For this reason, the input to SRR is usually preregistered using the well-known ICP algorithm.

Figure 8.
Comparison of point clouds and trajectories after application of different SLAM algorithms. Upper row: Birds-eye sliced view of the point clouds. Lower row: Side-view of the point clouds, aligned with the initial orientation. From left to right: (1) initial point cloud and trajectory. (2) after ICP, metascan variant, available in 3DTK—The 3D toolkit [34]. (3) after the first stage of the proposed method.

The SRR preregistration uses a highly precise metascan implementation, available from 3DTK—The 3D Toolkit [34]. However, the polygon-based approach outperformes ICP, which is illustrated in **Figure 8**. In the images of the left column, one sees a birds-eye view of the input to both algorithms. The center images show the resulting point cloud and trajectory after ICP. Due to the given trajectory, the algorithm is not able to establish meaningful point-to-point matches on the dataset, thus the output is no longer planar. The first stage we present in this work is shown in the images in the right column. Although some walls are not algined, the result is more planar and resembles the environment better than the output of ICP. Using this method before applying SRR leads to a faster convergence and more accurate solution in the second stage. In the next section, we analyze the accuracy of the resulting maps qualitatively, as well as quantitatively using high-precise ground truth point clouds.

4. Experiments

In this section, we describe the system setup and procedure of four experiments, which demonstrate unconventional trajectories. Note that all datasets are available at http://kos.informatik.uni-osnabrueck.de/3Dscans/. We compare three systems to each other that have been tested in the same environment. One other system had to be tested in a different building, which allowed for a long descent from a crane, which we therefore analyze separately. For our experiments, we use three kinds of motion: rolling on the ground, moving forward while swinging, and descending while rotating. All motion profiles were shown previously in **Figure 2a**. A birds-eye view of the environments in which the experiments were carried out is illustrated in **Figure 9**. In the left image, a square hallway can be seen which we used for the pendulum and rolling experiments, whereas the right image shows the firefighter school which we used for the descending experiment. We use the same LiDAR sensor in three experiments: the Livox Mid-100. It produces 300.000 points per second using three beams that scan in a non-repetitive, flower-shaped fashion, thus point density increases over

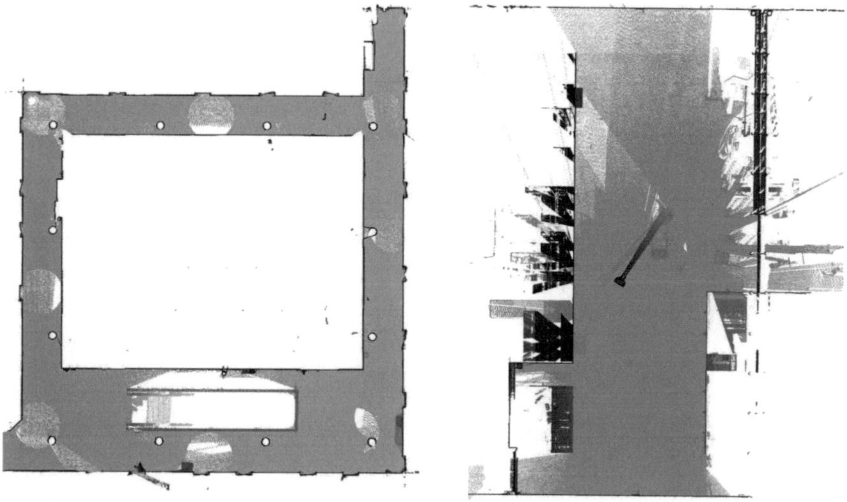

Figure 9.
Birds-eye view of the ground truth point clouds, acquired with a terrestrial laser scanner (TLS). The ceiling has been cropped for a better view. Left: Ground truth point cloud of an office hallway used for the rolling and pendulum experiments. Right: Ground truth point cloud of the state firefighters school in Würzburg, used for the descending experiment.

time. Each beam has a circular field of view (FOV) of 38.4°. Thus, three beams aligned in a row create a vertical FOV of 38.4° and horizontal FOV of 98.4°. The precision at 20-meter scanning distance is 2 cm and the angular accuracy is 0.1°. Further, the minimum scanning distance of the laser scanner is 1 m and the output frequency is set to 10 Hz. The maximum output frequency of the sensor is 50 Hz, yet point density then decreases. In future work, though, we want to test the systems also with 50 Hz LiDAR frequency. For all setups, a ROS installation on a Raspberry Pi 4 is used for onboard controlling and recording data. Additionally, we apply the rolling motion to a SICK LMS141 2D laser scanner with a maximum range of 40 m that operates at a scanning frequency of 50 Hz and resolution of 0.5°. Here a Raspberry Pi 3 is used for data collection. Inertial measurements are performed by PhidgetSpatial Precision 3/3/3 IMUs. The post-processing is performed after the experiments on a separate server.

4.1 Pendulum

As to the pendulum setup the system is equipped with an Intel T265 tracking camera, which uses a combination of feature tracking and internal IMUs to estimate its pose. The T265 unfortunately has been discontinued by Intel and is only available on the secondary market, which increased its price. As a budget alternative, we consider the Intel T261, which is still available, or performing visual-inertial-odometry manually, e.g. Intels RealSense-SLAM or VINS-Fusion [35] with Intels D435i. To test this setup in the hallways of an office-like environment (cf. upper left image of **Figure 2b**), we put the sensors inside a trailer net and swing them back and forth while walking. The movement itself consists of de- and accelerations to the front and back and slight movements up and down. Depending on the walking speed of the person, even an overall negative velocity of the sensor is possible if the pendulum swings faster backward than the person moves forward. Note that the view of the tracking camera is partially obscured by the trailer net, reflections off the shell, and the thread of the shell. As the camera and laser scanner are mounted near to each other with the same orientation, we assume that their coordinate systems coincide and thus, use no external calibration between the two. The duration of this motion was 281 s and covered a total distance of approx. 162 m (only walking, oscillation not included).

4.2 Rolling on flat surfaces

In this experiment, we put the sensors inside a spherical plastic shell and roll it on a flat surface manually (cf. bottom left image of **Figure 2b**). This time there is no tracking camera included. Rather, rotation, as well as position, estimates come from a combination of three Phidget 3/3/31044-0b IMUs. We use an IMU filter that is specialized for pose estimation of spherical robots [4]. The rolling duration was 691 s and covered the same distance as before of 162 m. Further, a similar experiment has been executed in the same environment and with the same trajectory, with the RADLER system as described in [8]. RADLER is a modified unicycle where the 2D laser scanner is mounted with its scanning plane parallel to the wheel axle thus creating a radial scanning pattern while rotating.

4.3 Crane descending

Unlike the previous experiments, this one was executed in a different environment that allowed for a long descent, i.e., in the building of the state firefighters school in Würzburg. We connected the robot to an outsourced processing machine via a 50 m tear-resistant tether cable (Fathom ROV Tether by BlueRobotics) which was rolled around a coil to perform the descending and ascending movement (cf. right image of **Figure 2b**). In this experiment, the sensor unrestrictedly rotates around the descending axis, corresponding to the cable direction. Note that the rotation itself is induced by the internal twist of the cable, not by any actuators. A spin encoder estimates the position, as it measures the rotation of the coil which directly corresponds to the height of the robot according to the helix arc length formula. The descent of the sphere covered a distance of 22 m and was performed within a duration of 402 s.

5. Evaluation

The resulting point clouds after application of the presented SLAM approach are now analyzed in terms of their accuracy, which we do by matching them against high-precise ground-truth models of the environment, given by a RIEGL VZ400 terrestrial laser scanner (TLS). It has an angular resolution of 0.08° and an accuracy of 5 mm. For the evaluation, we consider histograms that show a distribution of point-to-point errors. To create such histograms, we match the resulting point clouds against the corresponding ground truth maps, using ICP from 3DTK [34]. Then, we create a three-dimensional difference image by measuring all point-to-point errors. When calculating the difference images for any dataset, we use an octree-based filter where the voxels are smaller than 10 cm with a maximum of 5 points. Thus, the resolution of the different images is the same, i.e., histograms do not depend on point density anymore and are comparable to each other as long as they have been created for the same environment. **Figure 10** contains three histograms that were created in the office hallway environment. Broadly speaking, a distribution is better if its tail is short and its peak is located toward the left, i.e., zero point-to-point error. In particular, the

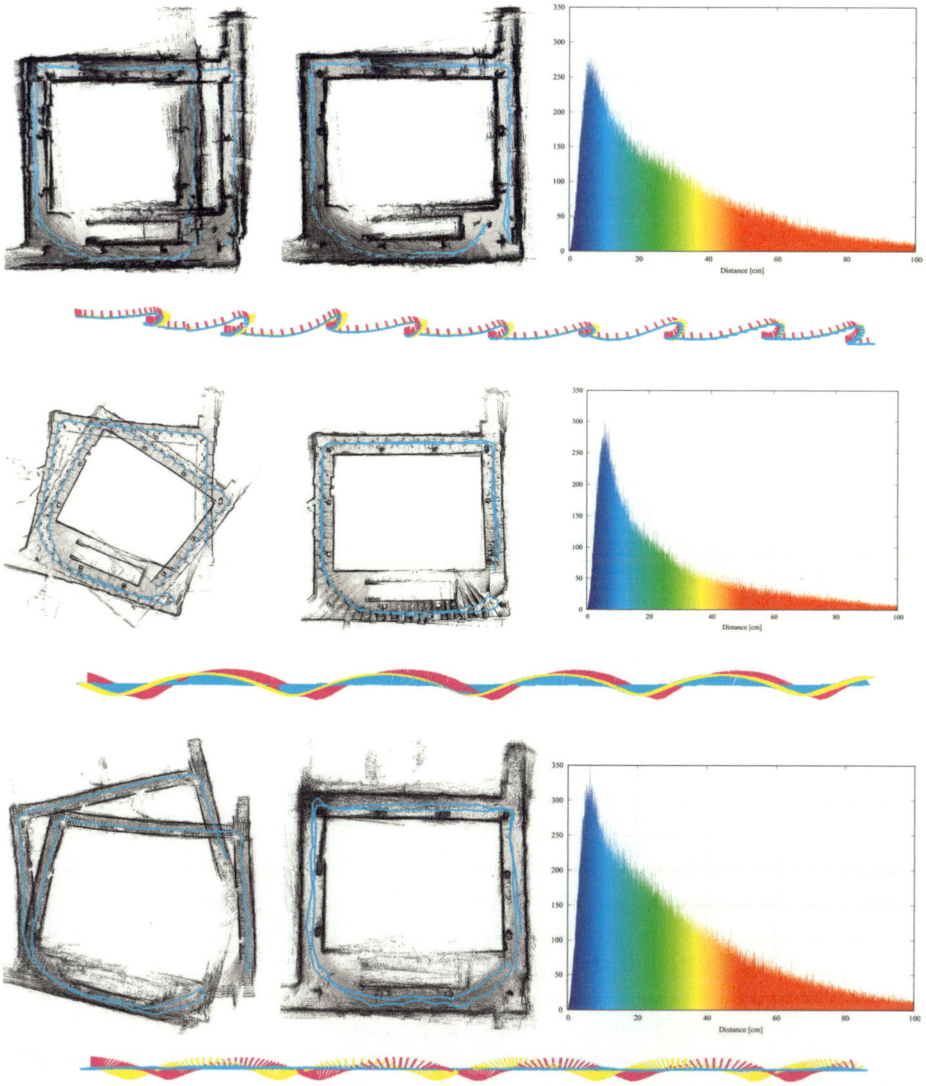

Figure 10.
Birds-eye view of point clouds acquired with the different setups. The ceiling has been cropped for a better view. A profile view showing sensor poses after registration, movement from left to right is also shown. Left column: Unprocessed point cloud with initial pose estimates, from IMUs and tracking camera (pendulum), IMUs and wheel encoder (RADLER) and IMUs only (spherical). Center column: Post-processed point cloud after application of proposed SLAM algorithm. Right column: Histograms showing the occurrences of certain point-to-point errors from the compared post-processed 3D point clouds to the reference ground truth point cloud. The colors denote distance, where blue corresponds to zero distance and red corresponds to 50 cm distance or more. (a): Pendulum setup. Mean point-to-point error is 28.31 cm with peak at 10.94 cm. (b): Spherical system. Mean point-to-point error is 28.70 cm with peak at 11.64 cm. (c): RADLER. Mean point-to-point error is 24.48 cm with peak at 12.10 cm.

quantity of most interest is the distribution mean, as it tells you the average point-to-point error. The individual histograms correspond to the pendulum, spherical system, and RADLER results, respectively. From the naked eye, one might suspect that RADLER performed best, while the spherical system performed worst. The following sections confirm this statement quantitatively and give a more detailed interpretation of the results.

5.1 Pendulum

Figure 10a shows that the initial pose estimations using the Intel T265 tracking camera is the most accurate, when compared to the methods without feature tracking (cf. left image of **Figure 10b** and **c**). In the other datasets, the IMUs struggle with yaw angle estimations especially at the corners, whereas here the feature tracking compensates for that. After registration, the map represents the environment better as before since there are no duplicate corridors left in the point cloud, and the optimized poses are consistent with the map. Yet the result is not perfect, e.g., the walls appear thick due to the large amount of motion distortion, the pillars are not perfectly aligned, and a sizeable duplicate wall remains uncorrected. Note that this is presumably because there is no external calibration between the tracking camera and laser scanner. Employing such is a task for future work and potentially increases the mapping accuracy. According to the mean point-to-point error ($\overline{E}$) from the histograms in **Figure 10**, this result ($\overline{E} = 28.31$ cm) resembles ground truth better than the rolling system ($\overline{E} = 28.70$ cm), but worse than RADLER ($\overline{E} = 24.48$ cm).

5.2 Rolling on flat surfaces

The following sections sum up the results for the mobile systems with rolling sensor trajectories, i.e., RADLER and the spherical system. As mentioned above, RADLER has the best similarity to ground truth according to **Figure 10**, whereas the spherical system has the worst.

5.2.1 2D LiDAR

Figure 10b presents the results of the experiment with RADLER, which were carried out in the same environment as before. The left image shows that the initial pose estimates have significantly more drift compared to the pendulum system, especially regarding the yaw angle. However, the walls appear thinner, and there is overall less noise due to the missing shell. After our SLAM, there are a few spots where the walls are not perfectly aligned, and a lot of noise remains between the walls. In comparison to the maps created with the pendulum and spherical system, though, RADLERs result resembles ground truth best. We suppose that this is because RADLER's 2D scanner operates at a higher frequency (50 Hz) and uses the rotational encoder in addition to the IMU for determining the systems pitch. The Livox Mid-100 scanner used for the other experiments operates on only 10 Hz, which is why fast trajectories lead to a more obscure scan and, thus thicker walls. Further, the 2D scanner from SICK is optimized for short-range measurements, whereas the 3D scanner from Livox is for long medium- to long-range measurements.

5.2.2 3D LiDAR

Figure 10c shows the results of the experiment with the spherical system. The initially estimated trajectory distance is the largest when compared against the other datasets (211.77 m compared to RADLER: 141.01 m, and pendulum: 148.30 m), indicating an overestimated radius parameter in the pose estimation model [4]. The resulting map resembles the actual scale of the environment better, the pillars are well aligned, and the corrected poses are consistent with the map. However, there are also

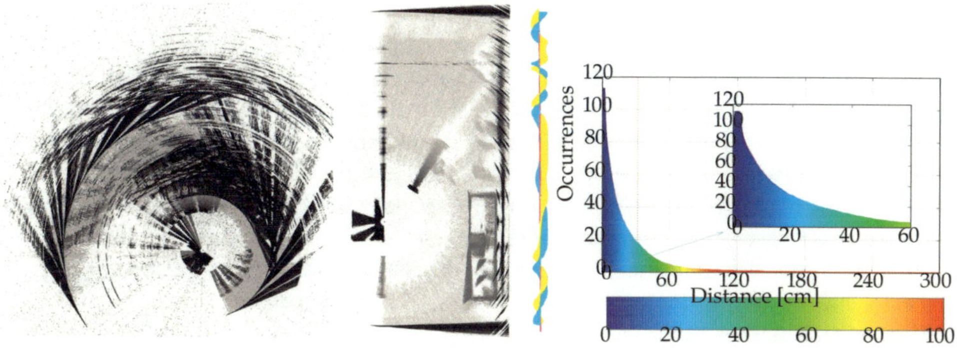

Figure 11.
In the images of 3D point clouds, the ceiling has been cropped for a better view. From left to right: (1) birds-eye view of the resulting 3D point cloud, acquired with the descending system using a spin-encoder and IMUs for pose estimation. (2) birds-eye view of the post-processed 3D point cloud. (3) profile view of the mobile systems pose, movement from top to bottom. (4) histogram showing the occurrences of certain point-to-point errors to ground truth. The colors denote distance, where blue corresponds to zero distance and red corresponds to 1 m distance or more. Mean point-to-point error is 31.6 cm with peak at 3.60 cm.

duplicate walls, remaining outliers, and noise due to remaining registration errors. According to the histograms in **Figure 10**, the resulting map has the least similarity to ground truth when compared to the pendulum and RADLER, which is consistent with previous observations. Note that in this dataset, the shell is attached to the sphere, which makes range measurements more noisy and adds outliers due to reflections off the shell.

5.3 Crane descent

This section presents the results for the crane descent experiment, which was conducted in a different environment than the previously mentioned results. Thus, the histogram is not really comparable to the ones in **Figure 10**, although it uses the same voxel filter to create the distance image. We still analyze the shape and mean point-to-point error of the distribution and to interpret them. The upper half of **Figure 11** shows a birds-eye view of the 3D point clouds in the same fashion as before. Note that the initial pose estimates are especially erroneous in one rotational dimension. This is the yaw rotation, which is especially difficult to detect for IMUs without the use of a magnetometer. As this experiment originated in the context of a space mission, using the magnetometer for inertial measurements was not an option. In the first stage of an out SLAM algorithm, we locked each other dimension but yaw from being optimized. The resulting map resembles the environment well, yet error remains due to low scanning frequency and motion distortion. The mean point-to-point error when comparing against ground truth (cf. right image of **Figure 9**) is 31.6 cm. However, the peak of the historgram is located at 3.60 cm, indicating that there is room for further improvement. We seek to improve on these results by accounting motion distortion and further reducing IMU drift in future work.

6. Conclusion

We have shown in this work that unconventional trajectories still pose problems for current SLAM algorithms, especially when using low FoV LiDARs. We built a flexible SLAM approach that shows the capabilities to register unconventional trajectories with large-scale pose estimation errors reliably. Further, we tested our SLAM system with three different unconventional movements: rolling, pendulum, and rotating crane descend. According to the previous accuracy evaluation, rolling on the floor is the most difficult scenario. The spherical shell of the system adds noise and outliers to the range measurement. Additionally, low overlap and sometimes no overlap make scan matching hard, even using polygons. Therefore, the success of SLAM using this trajectory type, compared with the other scanning methods, relies the most on the initial pose estimations. Moreover, this scenario has the most difficult initial pose estimation, due to the large accumulation of errors both in translation and rotation, which makes SLAM especially difficult. RADLER seems to have the best results regarding accuracy. This is because of the higher scanning frequency compared to the other experiments, but also because the rotational encoder on the wheel helps a lot with position estimation when compared to the spherical setup, which relies on constrained IMU integration. Therefore, it is sufficient to compensate mostly the accumulated rotational error via SLAM. Descending from the crane shows similar behavior: the rotational encoder on the cable reel makes position estimation fairly easy. Further, there are only negligible rotations in two principal axes. However, the faster and uncontrolled rotation around the cable leads to a much larger error in the corresponding axis of revolution, as well as to larger motion distortion. Since pose errors mostly accumulate in one rotational degree of freedom, our SLAM is still able to correct these via constrained optimization in the first stage. The pendulum setup, on the other hand, shows almost no rotational error in the initial pose estimations, because the visual-inertial odometry (VIO) of the Intel T265 camera compensates for IMU drift. VIO works reliably although the view of the camera is partially obscured by the trailer net, the thread of the shell, and reflections from the shell. Yet the relocalization module of the camera, which uses an internal feature map, fails once, as the visual features of the hallways are ambiguous. This leads to large positional errors in the initial pose estimates at the end of the trajectory. Our SLAM is able to correct these errors using the polygon-based optimization in the first stage. However, a lot of work remains to be done. In particular, we need to address the large drift of the IMUs and employ a more accurate external calibration, to improve the initial pose estimates even more. Furthermore, we aim to improve the quality of the maps by revisiting the global polygon model of the algorithms first stage and making it more dynamic. Moreover, we want to mitigate the effects of motion distortion in future work. Finally, the pose estimations need to be evaluated in terms of the achieved positional and rotational errors, e.g. with an external optical tracking system. Nevertheless, we provide significant insight and datasets with ground truth maps and are excited to contribute more to this rather unexplored field of research in future work.

Author details

Fabian Arzberger[1*], Jasper Zevering[1], Anton Bredenbeck[2], Dorit Borrmann[1] and Andreas Nüchter[1]

1 Robotics and Telematics, Julius Maximilian University of Würzburg, Würzburg, Germany

2 Delft University of Technology, Delft, The Netherlands

*Address all correspondence to: fabian.arzberger@uni-wuerzburg.de

References

[1] Li K, Li M, Hanebeck UD. Towards high-performance solid-state-LiDAR-inertial odometry and mapping. IEEE Robotics and Automation Letters. 2021; **6**(3):5167-5174

[2] Tan W, Zhang D, Ma L, Wang L, Qin N, Chen Y, et al. Semantic segmentation of UAV lidar point clouds of a stack interchange with deep neural networks. In: 2021 IEEE International Geoscience and Remote Sensing Symposium IGARSS. Brussels, Belgium: IGARSS; 2021. pp. 582-858

[3] Lin J, Zhang F. Loam livox: A fast, robust, high-precision LiDAR odometry and mapping package for LiDARs of small FoV. In: 2020 IEEE International Conference on Robotics and Automation. (ICRA). 2020. pp. 3126-3131. DOI: 10.1109/ICRA40945.2020.9197440

[4] Zevering J, Bredenbeck A, Arzberger F, Borrmann D, Nuechter A. IMU-based pose-estimation for spherical robots with limited resources. In: 2021 IEEE International Conference on Multisensor Fusion and Integration for Intelligent Systems. Karlsruhe, Germany: MFI; 2021. pp. 1-8

[5] Arzberger F, Bredenbeck A, Zevering J, Borrmann D, NÃ¼chter A. Towards spherical robots for mobile mapping in human made environments. ISPRS Open Jour Photogrammetry and Remote Sensing. 2021;**1**:100004

[6] Borrmann D, Elseberg J, Lingemann K, Nüchter A. The 3D Hough transform for plane detection in point clouds: A review and a new accumulator design. 3D Research. 2011;**2**:1-13

[7] Elseberg J, Borrmann D, Nuchter A. 6DOF Semi-rigid SLAM for Mobile Scanning. In: 2012 IEEE/RSJ International Conference on Intelligent Robots and Systems. Vilamoura-Algarve, Portugal: IROS; 2012. pp. 1865-1870

[8] Borrmann D, Jörissen S, Nuchter A. Radler – A RADial LasER scanning device. In: Proc. of the Int. Symp. on Experimental Research. Buenos Aires, Argentina; 2020. pp. 655-664

[9] Zevering J, Bredenbeck A, Arzberger F, Borrmann D, Nüchter A. LUNA-A laser-mapping unidirectional navigation actuator. In: International Symposium on Experimental Robotics. Valletta, Malta: ISER; 2020. pp. 85-94

[10] Rossi AP, Maurelli F, Dreger H, Mathewos K, Pradhan N, Pozzobon R, et al. DAEDALUS - Descent and Exploration in Deep Autonomy of Lava Underground Structures. JMU Würzburg: Inst.für Informatik; 2021. p. 21

[11] ESA. ESA plans mission to explore lunar caves; 2021. Available from: https://www.esa.int/Enabling_Support/Preparing_for_the_Future/Discovery_and_Preparation/ESA_plans_mission_to_explore_lunar_caves

[12] Bosse M, Zlot R, Flick P. Zebedee: Design of a Spring-Mounted 3-D range sensor with application to Mobile mapping. IEEE Transactions on Robotics. 2012;**28**(5):1104-1119

[13] Lehtola VV, Virtanen JP, Vaaja MT, Hyyppä H, Nüchter A. Localization of a Mobile laser scanner via dimensional reduction. ISPRS Journal of Photogrammetry and Remote Sensing (JPRS). 2016;**121**:48-59

[14] Geoslam. ZEB Revo RT;. Accessed: February 24, 2022. https://www.geoslam.com/solutions/zeb-revo-rt/

[15] Leica. Leica BLK2GO;. Accessed: February 24, 2022. https://shop.leica-geosystems.com/global/blk2go-overview

[16] Alismail H, Browning B. Automatic calibration of spinning actuated Lidar internal parameters. Jour Field Robotics. 2015

[17] Besl PJ, McKay ND. A method for registration of 3-D shapes. IEEE Transactions on Pattern Analysis and Machine Intelligence. 1992;**14**(2): 239-256

[18] Lu F, Milios EE. Globally consistent range scan alignment for environment mapping. Auton Robots. 1997;**4**:333-349

[19] Borrmann D, Elseberg J, Lingemann K, Nüchter A, Hertzberg J. Globally consistent 3D mapping with scan matching. Robotics and Auton Systems. 2008;**56**(2):130-142

[20] Gentil CL, Vidal-Calleja T, Huang S. IN2LAMA: INertial Lidar Localisation and MApping. In: 2019 International Conference on Robotics and Automation. Montreal, QC, Canada: ICRA; 2019. pp. 6388-6394

[21] Droeschel D, Behnke S. Efficient Continuous-Time SLAM for 3D Lidar-Based Online Mapping. In: IEEE Int Conf on Robotics and Automation. Brisbane, Australia: ICRA; 2018

[22] Förstner W, Khoshelham K. Efficient and accurate registration of point clouds with plane to plane correspondences. In: 2017 IEEE Int. Conf. on Computer Vision Workshops. Venice, Italy: ICCVW; 2017. pp. 2165-2173

[23] Grant WS, Voorhies RC, Itti L. Efficient Velodyne SLAM with point and plane features. Auton Robots. 2019;**43**: 1207-1224

[24] Geneva P, Eckenhoff K, Yang Y, Huang G. LIPS: LiDAR-Inertial 3D Plane SLAM. In: Proc. of the IEEE/RSJ Int. Conf. on Intelligent Robots and Systems (IROS '18). 2018. pp. 123-130

[25] Zhou L, Wang S, Kaess M. LSAM: LiDAR smoothing and mapping with Planes. In: Proc. of IEEE Int. Conf. On Robotics and Automation (ICRA '21). Xi'an, China; 2021

[26] Wei X, Lv J, Sun J, Pu S. Ground-SLAM: Ground Constrained LiDAR SLAM for Structured Multi-Floor Environments; 2021. DOI: 10.48550/ARXIV.2103.03713

[27] Favre K, Pressigout M, Marchand E, Morin L. A plane-based approach for indoor point clouds registration. In: ICPR 2020 - 25th Int. Conf. on Pattern Recognition. Milan, Italy; 2021

[28] Taguchi Y, Jian YD, Ramalingam S, Feng C. Point-plane SLAM for hand-held 3D sensors. In: 2013 IEEE International Conference on Robotics and Automation. Karlsruhe, Germany: IEEE; 2013. pp. 5182-5289

[29] LIVOX. Livox mapping;. Accessed June 30, 2022. https://github.com/Livox-SDK/livox_mapping

[30] Zhang J, Singh S. LOAM: Lidar odometry and mapping in real-time. In: Robotics: Science and Systems Conf. Berkeley, California. 2014

[31] Zeiler MD. ADADELTA: An adaptive learning rate method. CoRR. 2012

[32] Arun KS, Huang TS, Blostein SD. Least-squares fitting of two 3-D point sets. IEEE Transactions on Pattern Analysis and Machine Intelligence. 1987; **9**(5):698-700

[33] Golub G, Kahan W. Calculating the singular values and Pseudo-inverse of a matrix. Journal of the Society for Industrial and Applied Mathematics: Series B, Numerical Analysis. 1965;**2**(2): 205-224

[34] Nüchter A, Lingemann K. 3DTK—The 3D Toolkit. 2011. https://slam6d.sourceforge.io/index.html

[35] Qin T, Cao S, Pan J, Shen S. A General Optimization-based Framework for Global Pose Estimation with Multiple Sensors; 2019

Section 2

Key Software Components

Chapter 3

Autonomous Mobile Mapping Robots: Key Software Components

Janusz Będkowski and Jacek Szklarski

Abstract

This chapter discusses key software components of autonomous mobile mapping robots equipped with an inertial measurement unit (IMU) and light detection and ranging (LiDAR). In recent years, new LiDARs with nonrepetitive scanning pattern have appeared in the market. They are also equipped with an IMU; thus, the front end of simultaneous localization and mapping (SLAM)—a robust LiDAR-inertial odometry framework—significantly improves unmanned ground vehicles (UGVs) and unmanned aerial vehicles (UAV) in 3D mapping scenarios. Our study incorporates FAST-LIO as the front end of SLAM. The main focus is a lightweight back-end implementation of pose graph simultaneous localization and mapping (SLAM). It is an alternative solution to state-of-the-art g2o or GTSAM implementations. We also elaborate on iterative closest point, normal distributions transform, and their extension for multiview 3D data registration/refinement. It is based on C++ using Eigen library. This chapter also discusses path planning in already mapped environment. All software components are available as open-source projects.

Keywords: multiview normal distributions transform, SLAM, path planning, coverage

1. Introduction

This chapter presents key software components for autonomous mobile mapping robots shown in **Figure 1** (software components are available in [1, 2]). This set of consecutive functionalities is composed of robust light detection and ranging (LiDAR)-inertial odometry FAST-LIO [3], pairwise matching algorithm (iterative closest point [ICP] or normal distributions transform [NDT]) [1] for minimizing an error for loop closures, pose graph simultaneous localization and mapping (SLAM) [2], final refinement (multiview NDT) [1], and path planning. These functionalities are the core components for autonomous mobile mapping robots equipped with an inertial measurement unit (IMU) and LiDAR.

Autonomous mobile mapping robots have already been widely investigated within the context of commercial applications, for example, power line inspection [4], smart factory production [5], offshore oil plant [6], and nuclear power plant (NPP) inspection [7]. Robots improve rescue missions in hazardous environments [8]. The research related to COVID-19 and public support for autonomous technologies shows great interest in artificial intelligence (AI) direction [9]. There are plenty of areas for

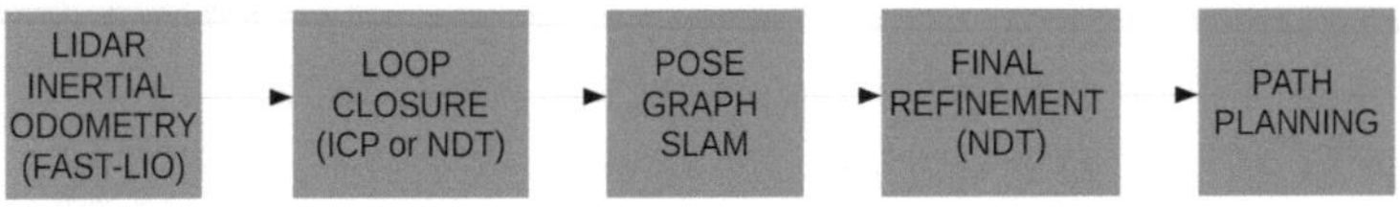

Figure 1.
Scheme of key software components elaborated in this chapter.

autonomous mobile mapping robots such as floor scrubbing, delivery, warehouse, and service robots. The rapid improvements in autonomous mobile mapping are evident for LiDAR with a nonrepetitive scanning pattern [10]. This LiDAR is capable of acquiring massive 3D data in short time with a limited field of view. The advantage is the long range, even up to 500 meters and high density of points covering entire measurement cone in short time. Narrow field of view can be extended by multiple LiDAR systems [11]. Owing to synchronized IMU data and robust feature classification, a robust LiDAR-inertial odometry framework significantly improves unmanned ground vehicles (UGVs) and unmanned aerial vehicles (UAVs) in 3D mapping scenarios [3, 12]. Moreover, the overall cost of such LiDAR is rather small compared with its competition. It is advised to study [13] for further precision and accuracy comparative evaluation.

SLAM is a core component of the autonomous mobile mapping robot that builds a map based on the estimated trajectory and estimates this set of consecutive poses based on this map [14]. SLAM is composed of front-end capable reconstructing smooth and continuous trajectory from onboard sensors. This trajectory is affected by constantly growing error. To reduce this error, the additional measurements should be incorporated into the back end as the so-called loop closure. The back end is typically solved using the pose graph SLAM method implemented in g2o [15] and GTSAM [16] frameworks. An alternative implementation is available in [2] that extends the possibility of rotation matrix parameterizations. Pose graph SLAM optimizes a graph composed of vertices (poses) and edges (consecutive odometry readings, loop closures, and other constraints); thus, it is supposed to preserve the shape of an initial trajectory (motion model) and minimize an error between observed (current relative pose) and measured (desired relative pose) loop closure edges. For the calculation of desired relative pose between two scans, an iterative closest point [17] or normal distributions transform [18] can be incorporated. The final step of the 3D mapping can be the final refinement of all 3D measurements performed with, for example, multiview normal distributions transform [19]. A similar approach is evident in general mobile mapping applications [20, 21].

The last functionality elaborated in this chapter is path planning being a fundamental software component of the autonomous mobile mapping system dedicated for missions where full coverage is desirable [22]. Examples are nuclear power plan inspection [23] and cleaning robotics [24]. In order to perform a mapping task, a mobile robot has to act according to some kind of a motion planning algorithm. This is also true for other related tasks such as inspecting, searching, cleaning, image mosaicking, etc. There are many factors that determine which algorithm should be used for the path plan generation:

- Is the map of the environment known in advance?
- What are the available sensors (LiDARs, RGB(D) cameras, proximity sensors, etc)?

- How many robots participate in the task? If more than one, what are the means of communication between units?
- How the environment map is represented? Is the map 2D or 3D? Does robot move on a planar space or is it a UAV (six degrees of freedom)?
- What are the available computational resources for path calculations and what is the required working regime: real-time online or offline planning?
- What is the ratio of covering radius to the size of the robot?

All the aspects mentioned above profoundly impact the algorithms necessary to guide the mobile robot position. For example, if the map of the environment is not known in advance, one should focus on a version of SLAM with exploration algorithm. If, on the other hand, the map is known and offline planning is allowed, one may use a solver that generates plan giving some near-optimal path plan. The optimal criteria can also vary depending on a specific application: minimization of coverage time, minimization of energy, prioritization of certain regions, etc. If there is a group of robots involved, the question of equal workload distribution must also be addressed.

A path planning algorithm should also take into account kinematic properties of robots involved in the process. A different algorithm will be applied for path generation for aircraft taking aerial images for mosaicking and a different one for a mobile robot cleaning floor in a warehouse.

For the autonomous mapping of unknown environments, in order to obtain a map of the environment, a robot should be able to simultaneously localize and map and, at the same time, explore the environment. Path planning, in this context, is related to the exploration process. The robot should gradually explore the environment, and a map is incrementally built for new poses, while the robot localizes itself in this map. New, temporary goal poses are often chosen by means of frontier extraction [25]. Frontier areas represent boundary regions between known (mapped) and unknown regions of the environment. Robot motion between its current position and such temporary goal is realized by a typical path planning, which navigates between waypoints while avoiding obstacles. Details regarding the frontier exploration vary depending on application and may also include exploration in 3D, for example, [26].

One of the fundamental applications of mobile robots is to perform a coverage task. Such tasks that the robot will visit points in the environment, eventually visiting (or observing) the entire region of interest. This is necessary for tasks like cleaning, mowing, harvesting, planting, spraying, mapping, searching, painting, mosaicking, etc. Normally, the first step for such application is to obtain the map, for example, by means of exploration with SLAM. If the map is known, and the robot is able to localize itself using the map, a coverage path planning (CPP) algorithm should be employed in order to find consecutive waypoints. The area of interest will be covered entirely after the robot will visit all the points. The problem of CPP is well known and well studied in the field of robotics [27, 28]. Nevertheless, it remains challenging and even the simplest variants, like the lawnmower problem, are NP-hard [29]. There exist a large number of exact, approximate, and heuristic algorithms to solve many variations of both types of CPP [30–32].

The rest of this chapter is organized as follows. Section 2 discusses key software components for SLAM. Section 3 addresses path planning in known environments.

Section 4 shows an example of mobile mapping applications. Finally, Section 5 concludes this chapter.

2. Key software components for SLAM

In this section, the key software components are elaborated. The fundamental element of SLAM is an observation equation. The set of optimization equations builds an optimization system. Finding an optimal solution results in the final map and trajectory. The proposed lightweight implementation uses symbolic computing in Python (SymPy) [33] to generate C++ code for each observation equations. An open-source project is available in [2].

2.1 Observation equations

Observation Eq. (1) is composed of a target value y_i, a model function $\Psi_{[\beta]}(\boldsymbol{x}_i)$, and its *residual* r_i defined as the difference between the target value and the value of the model function for $\boldsymbol{x}_i$ and state vector β

$$\boxed{\underbrace{r_i}_{residual} = \underbrace{y_i}_{target\ value} - \underbrace{\Psi_{[\beta]}(\boldsymbol{x}_i)}_{model\ function}} \tag{1}$$

where β is the vector of n optimized parameters. The weighted nonlinear least squares optimization method finds the optimal n parameter values (β) by minimizing the objective function being a sum of C squared residuals

$$Sum = \sum_{i=1}^{C} r_i^2 = \underbrace{\sum_{i=1}^{C} \left(y_i - \Psi_{[\beta]}(\boldsymbol{x}_i)\right)^2}_{objective\ function}. \tag{2}$$

Therefore, the optimization problem is defined as

$$\boxed{\beta^* = \min_{\beta} \sum_{i=1}^{C} \left(y_i - \Psi_{[\beta]}(\boldsymbol{x}_i)\right)^2} \tag{3}$$

where there are C observation equations. It is efficiently solved using the iterative Levenberg-Marquardt algorithm [34]. A single kth iteration provides an update for β given as

$$\beta^{k+1} = \beta^k + \left(\overset{\Psi}{\boldsymbol{J}}{}^{\top} \boldsymbol{W} \overset{\Psi}{\boldsymbol{J}} + \lambda \boldsymbol{I}\right)^{-1} \overset{\Psi}{\boldsymbol{J}}{}^{\top} \boldsymbol{W} \boldsymbol{r}(\beta^k) \tag{4}$$

where $\boldsymbol{I}$ is the identity matrix, $\overset{\Psi}{\boldsymbol{J}}$ is the Jacobian of the model function, and $\boldsymbol{W}$ is the weight matrix modeling the impact of the observation equation into the optimization process. λ starts from an initial small value. During the optimization process, λ increases once $Sum = \sum_{i=1}^{C} r_i^2$ decreases; otherwise, λ decreases and the optimization process starts from the previous step. The observation equation for pose graph SLAM is given as

$$\underbrace{\begin{bmatrix} t_x^\delta \\ t_y^\delta \\ t_z^\delta \\ \omega^\delta \\ \varphi^\delta \\ \kappa^\delta \end{bmatrix}}_{residuals} = \underbrace{\begin{bmatrix} t_x \\ t_y \\ t_z \\ \omega \\ \varphi \\ \kappa \end{bmatrix}}_{target\ values} - \underbrace{m2v_{[t_x,t_y,t_z,\omega,\varphi,\kappa]}\left([\boldsymbol{R},\boldsymbol{t}]^{12}\right)}_{model\ function} \tag{5}$$

where $\begin{bmatrix} t_x^\delta & t_y^\delta & t_z^\delta & \omega^\delta & \varphi^\delta & \kappa^\delta \end{bmatrix}^\top$ are *residuals*, $\begin{bmatrix} t_x & t_y & t_z & \omega & \varphi & \kappa \end{bmatrix}^\top$ are *target values*, and $m2v_{[\beta]}\left([\boldsymbol{R},\boldsymbol{t}]_{12}\right)$ is the *model function*. Target values describe the desired edge (relative pose expressed as translation (t_x, t_y, t_z) and orientation $(\omega, \varphi, \kappa)$ between two optimized vertices (poses) of the graph. Relative pose $[\boldsymbol{R},\boldsymbol{t}]_{12}$ from pose $[\boldsymbol{R},\boldsymbol{t}]_1$ to pose $[\boldsymbol{R},\boldsymbol{t}]_2$ is given as

$$\begin{bmatrix} \boldsymbol{R} & \boldsymbol{t} \\ \boldsymbol{0}^{1\times3} & 1 \end{bmatrix}_{12} = \left(\begin{bmatrix} \boldsymbol{R} & \boldsymbol{t} \\ \boldsymbol{0}^{1\times3} & 1 \end{bmatrix}_1\right)^{-1} \begin{bmatrix} \boldsymbol{R} & \boldsymbol{t} \\ \boldsymbol{0}^{1\times3} & 1 \end{bmatrix}_2. \tag{6}$$

Function $m2v_{[\beta]}\left([\boldsymbol{R},\boldsymbol{t}]_{12}\right)$ retrieves $\beta = (t_x, t_y, t_z, \omega, \varphi, \kappa)^\top$ for the Tait-Bryan parametrization of the rotation matrix. This parameterization is essential to preserve the orthonormality of the rotation matrix during the optimization process. It is important to notice that other parameterizations exist, such as quaternion, Rodrigues, etc., but this is not the main topic of this chapter.

The iterative closest point algorithm finds the relative pose between two point clouds by incorporating the following source-point-to-target-point observation equation:

$$\underbrace{\begin{bmatrix} x^\delta \\ y^\delta \\ z^\delta \end{bmatrix}}_{residuals} = \underbrace{\begin{bmatrix} x^{tg} \\ y^{tg} \\ z^{tg} \end{bmatrix}}_{target\ values} - \underbrace{\Psi_{[\boldsymbol{R},\boldsymbol{t}]}\left(\boldsymbol{R},\boldsymbol{t},x^l,y^l,z^l\right)}_{model\ function} \tag{7}$$

where, $\begin{bmatrix} x^\delta & y^\delta & z^\delta \end{bmatrix}^\top$ are *residuals*, $\begin{bmatrix} x^{tg} & y^{tg} & z^{tg} \end{bmatrix}^\top$ are *target values*, and $\Psi_{[\boldsymbol{R},\boldsymbol{t}]}\left(\boldsymbol{R},\boldsymbol{t},x^l,y^l,z^l\right)$ is the *model function* that transforms 3D points (x^l,y^l,z^l) expressed in the local coordinate system into the global one.

The normal distributions transform algorithm is an alternative solution for pairwise matching with ICP, and it can be easily extended for multiview point cloud data registration (final refinement of the 3D map). It decomposes the 3D scene into a regular grid where for each cell, the centroid $\boldsymbol{\mu}$ and the covariance Σ are calculated with formulas (8) and (9). These formulas incorporate all points $\boldsymbol{P}_k^g$ in a single cell expressed in the global coordinate system

$$\boldsymbol{\mu} = \frac{1}{m}\sum_{k=1}^{m} \boldsymbol{P}_k^g \tag{8}$$

$$\Sigma = \frac{1}{m-1}\sum_{k=1}^{m}\left(\boldsymbol{P}_k^g - \boldsymbol{\mu}\right)\left(\boldsymbol{P}_k^g - \boldsymbol{\mu}\right)^{\mathsf{T}}. \tag{9}$$

The NDT observation equation is given as

$$\boxed{\underbrace{\begin{bmatrix} x^\delta \\ y^\delta \\ z^\delta \end{bmatrix}}_{residuals} = \underbrace{\begin{bmatrix} \mu_x \\ \mu_y \\ \mu_z \end{bmatrix}}_{target\ values} - \underbrace{\Upsilon_{[\boldsymbol{R},\boldsymbol{t}]}\left(\boldsymbol{R}, \boldsymbol{t}, x^l, y^l, z^l\right)}_{model\ function}} \tag{10}$$

where the target value is $\boldsymbol{\mu}$ and $\Sigma^{-1} = \boldsymbol{W}$ for each NDT observation equation incorporated in the Levenberg–Marquardt algorithm from eq. (4).

3. Path planning in known environments

In this section, we will focus on a practical example of a cleaning robot whose task is to clean a large area. Therefore, one needs to apply a path planning algorithm for a single device that moves in a known environment, and the map of static obstacles is known in advance (c.f. [35]).

3.1 Map decomposition

A cleaning robot has the coverage area equal to the area of its cleaning/sweeping device. For industrial cleaners, it is a dedicated brush equipped with a water and soap reservoir. Consequently, the size of the coverage area, being a circle with radius r_{cov}, is comparable with the size of the robot. The area to be cleaned is a large warehouse with the total area A, so it should be assumed that $r_{\text{cov}} \ll A$. From this assumption, it follows that any grid-based approach for planning should be avoided. This is because in most grid-based methods, the time for finding a solution grows significantly with the grid size (for many methods even exponentially [28]). For the discussed problem, the number of grid cells would be too large to come up with a feasible solution.

It should be noted that this is not always the case for coverage problems. For example, photo mosaicking has much larger r_{cov} than the size of the robots, for example, UAVs equipped with cameras. Area being photographed is much larger than the area of the device. Such problems may use a different planning approach than the one for cleaning robots.

Consequently, a solution based on the geometric decomposition of the grid map into a set of polygons should be considered. Afterward, a path on this set of polygon is found in a way that minimizes a given optimization criterion, in this case the total coverage time. The pipeline for the system is depicted in **Figure 2**.

After mapping the environment with SLAM and the proper optimization, a grid map of static obstacles is obtained. There are a number of ways to convert such a grid map into a set of polygons. This process is known as map decomposition. The most well-known method of decomposition is the *trapezoidal decomposition* where the grid map is “scanned” line by line along one direction and trapezoids are found, which cover the free space completely. Afterward, trapezoids are covered by simple back-and-forth motion patterns.

DOI: http://dx.doi.org/10.5772/intechopen.110549

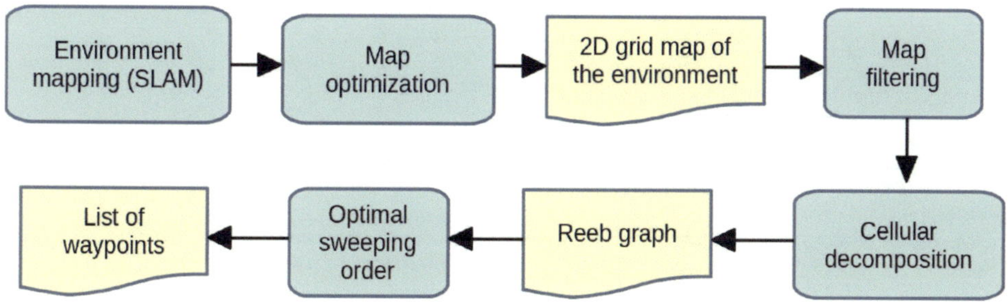

Figure 2.
A processing pipeline for the generation of a coverage path plan for a single robot operating in an environment that is first mapped with the SLAM method.

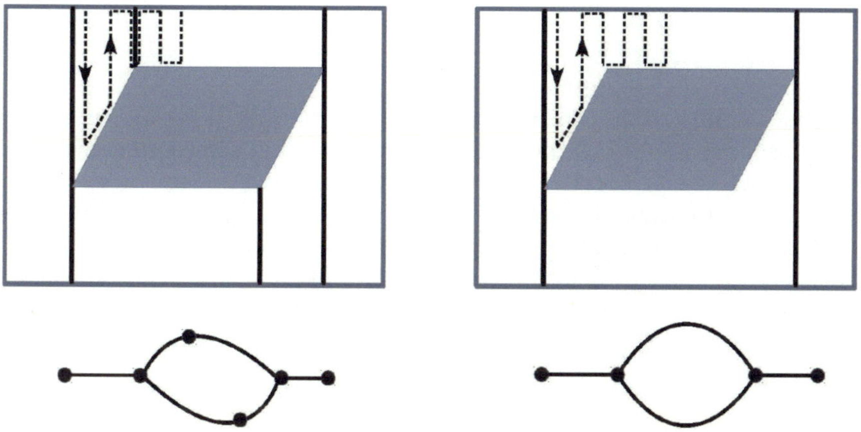

Figure 3.
An example of trapezoidal and boustrophedon decompositions into cells together with theirs Reeb graphs (bottom). The latter decomposition allows for better sweeping pattern fit (the dashed line).

The main drawback of this simple decomposition is the fact that it generates only convex polygons, and therefore, it results in a large number of polygons and suboptimal sweeping patterns (c.f. **Figure 3**). Another possibility is to apply the so-called *boustrophedon cellular decomposition* (BCD), which also generates nonconvex cells [36]. However, these polygons can also be covered only by zig-zag motions, usually in a more efficient way than for the trapezoidal decomposition. It should be noted that some more sophisticated decompositions have been proposed in the literature, for example, [37]. In such an approach, optimization is focused in the decomposition process itself. Here, however, we optimize the path by finding proper sweeping patterns at a later stage.

After the decomposition, a set of polygons is obtained. A geometrical relation between these polygons may be represented by the so-called Reeb graph. The Reeb graph is a special type of a graph for environment representation, in which each link corresponds to a polygon and each node represents an adjacency between the polygons. Consequently, the problem may be treated with the help of existing solutions known from graph theory. This can be done by formulating the optimization problem into a variant of the traveling salesman problem and employing some known efficient solvers.

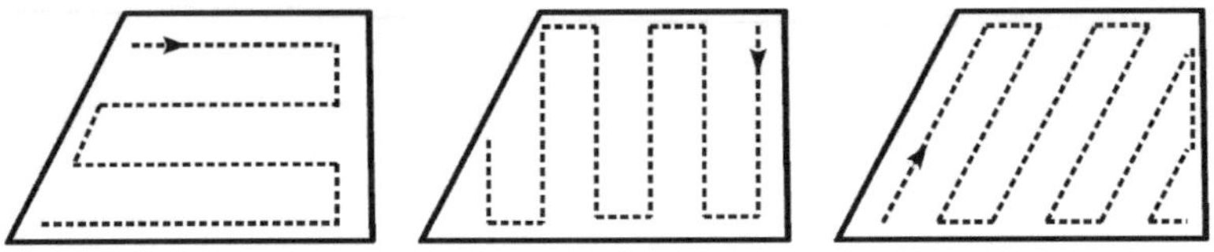

Figure 4.
For each cell, the coverage time is determined by sweeping direction and entry point into the cell (i.e., start and end vertices). The figure depicts some possible entry/exit points and directions for a sample trapezoid.

3.2 Finding near-optimal sweeping patterns

Let us consider the covering process for a single polygon. Usually, in order to minimize the coverage time, one needs to minimize the number of turns since, for each turn, a robot must slow down, stop, turn, and accelerate again to its maximum velocity. For a single polygon, various points of entrance for the zig-zag pattern should be considered (see **Figure 4**).

After the decomposition process, the entire environment is represented as a Reeb graph. The order of visiting all the links, that is, polygons, determines entry points to each of them and, therefore, the time cost associated with covering it. It can be shown that finding the minimum of the total time required for covering all the polygons is equivalent to solving the equality generalized traveling salesman problem (EG-TSP) [38]. This is an NP-hard problem for which approximate heuristic solvers may be applied. The results presented in this section are obtained with the use of a memetic solver, as proposed in [38].

4. Example applications

4.1 Robust LiDAR-inertial odometry and multiview NDT

Figure 5 demonstrates the result of FAST-LIO [3] as the 3D point cloud of underground garage recorded using Livox AVIA LiDAR. This robust LiDAR-inertial odometry provides an input for multiview NDT shown in **Figure 6**.

4.2 Pose graph SLAM

This section demonstrates the pose graph SLAM functionality available with data in [2]. **Figure 7** demonstrates the 2D case and **Figure 8** is related to the 3D case. Pose graph SLAM implementation efficiently solves the optimization problem represented as a consecutive set of poses (trajectory) connected via odometry readings (edges) and loop closure edges. This is a core component of the autonomous mobile mapping robot.

4.3 Final refinement with NDT

Multiview normal distributions transform 3D data registration is capable to increase the accuracy of the 3D map, as shown in **Figure 6**. The implementation is rather offline since it requires plenty of calculations. These calculations are related

Figure 5.
Result of robust LiDAR-inertial odometry FAST-LIO [3] as the 3D point cloud of underground garage.

mostly to 3D data decomposition, where for each 3D bucket, the mean value and covariance are calculated. This method is efficient mostly for urban environments with many planar shapes.

4.4 Path planning

As an example of a real-world application of a coverage task, let us consider the cleaning process of an underground garage. First, the garage is scanned with 3D laser scanners; it is optimized and flattened to a 2D obstacle map (see **Figure 9**, top). Afterward, the map is used for robot motion planning and navigation.

The result of boustrophedon decomposition for this map is shown in **Figure 9** (middle). In this particular case, it consists mostly of rectangles. The next stage is to find the covering path that connects all these polygons. In order to formally state the optimization goal, one needs to define kinematic characteristics of the robot. Here, we assume realistic parameters: $a_{\max} = 0.3\ \text{ms}^{-2}$ and $v_{\max} = 1\ \text{ms}^{-1}$. This corresponds to real devices that are being used for the cleaning tasks [35]. After using the memetic solver for the associated EG-TSP, a trajectory for the robot is obtained. It is shown at the bottom of the figure. For the depicted scale and the assumed kinematic model, the total time for coverage is 2302 s in this case.

In order to validate the approach for path planning and to estimate its usefulness, one should use a large number of maps, perform planning, and measure efficiency. One possibility is to use a synthetic albeit realistic set of layouts provided by Li et al. [39]. Based on real experiments with LiDARs, the authors have developed a method to generate about 60,000 various maps, which can be used by researchers to test various algorithms. Some example layouts together with the planned coverage paths are depicted in **Figure 10**.

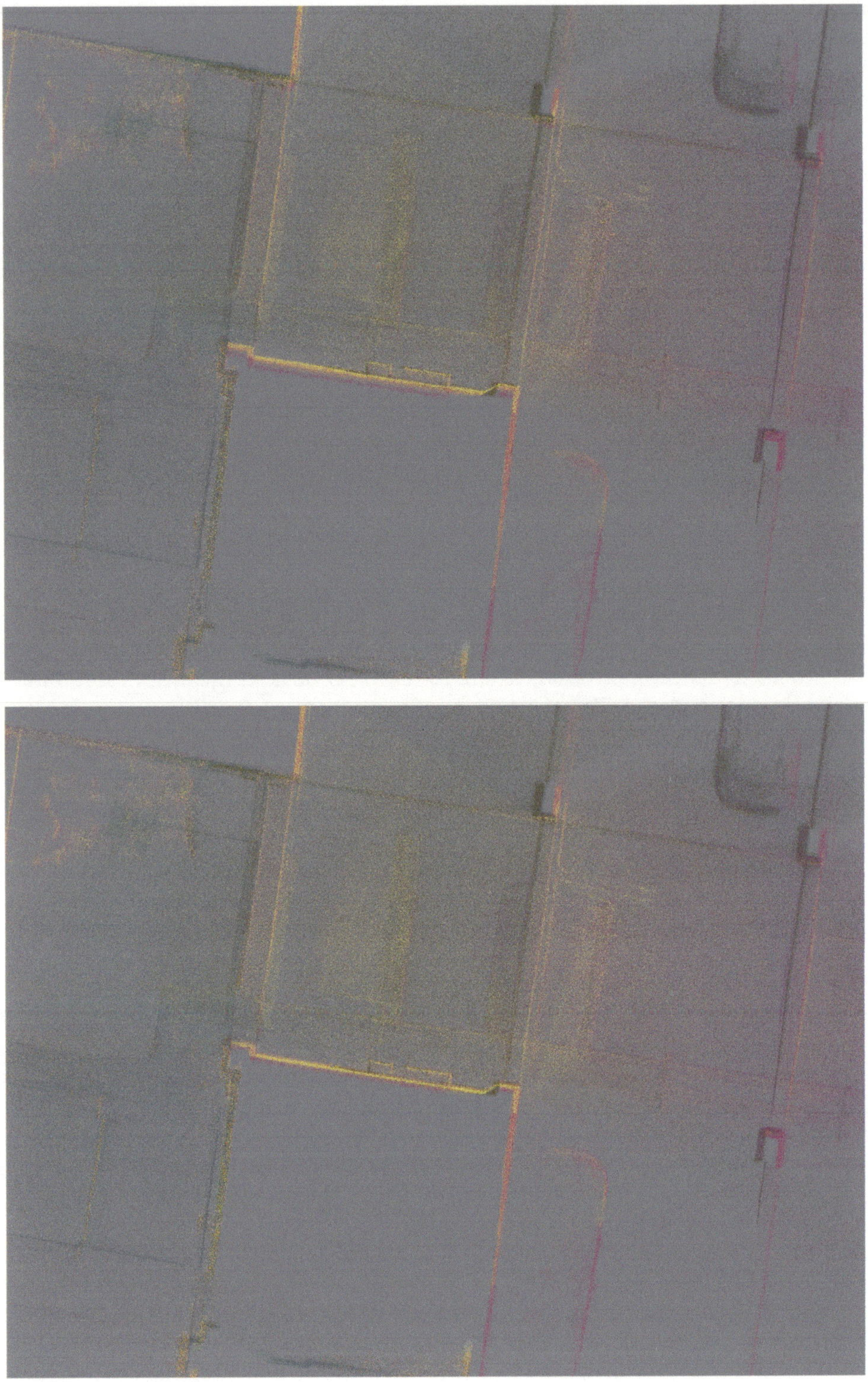

Figure 6.
Top: input data produced by robust LiDAR-inertial odometry FAST-LIO [3] from ***Figure 5****. Bottom: result of final refinement with multiview NDT.*

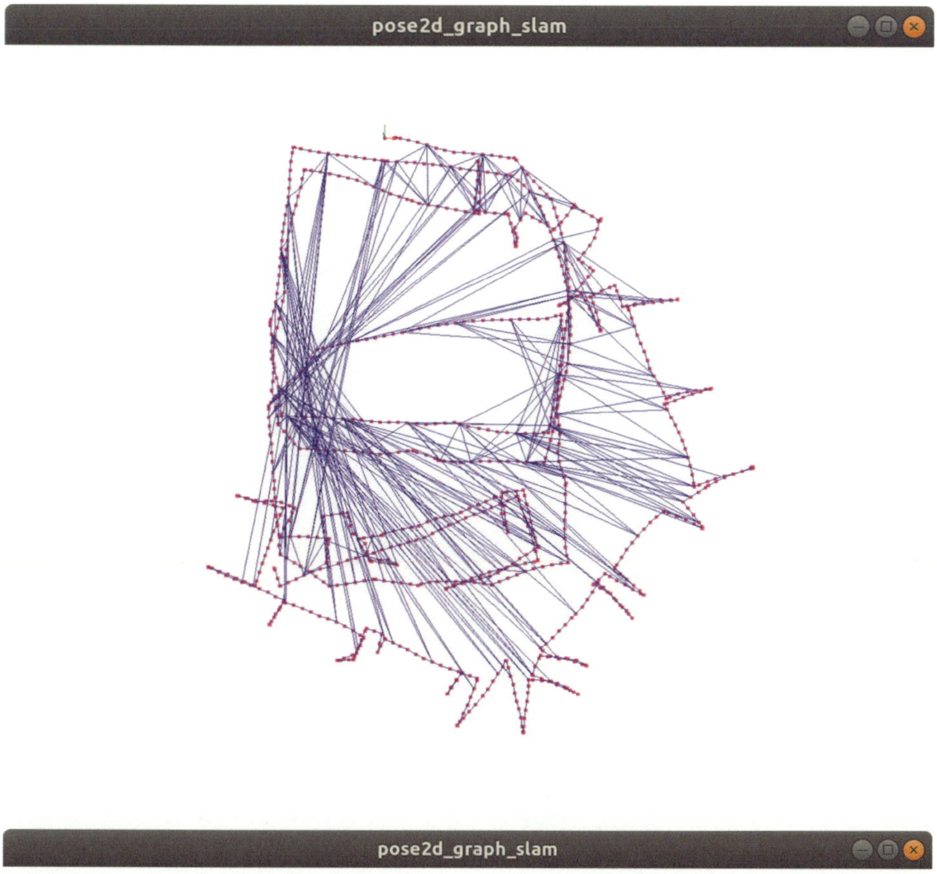

Figure 7.
Top: input data for pose graph SLAM (purple dots: graph vertices, blue lines: graph edges). Bottom: result of pose graph SLAM, 2D case.

Figure 8.
Top: input data for pose graph SLAM (purple dots: Graph vertices, blue lines: Graph edges). Bottom: result of pose graph SLAM, 3D case.

5. Conclusion

This chapter elaborates key software components of autonomous mobile mapping robots equipped with Livox AVIA LiDAR. It is new LiDAR with a nonrepetitive scanning pattern equipped also with the IMU. LiDAR and IMU are synchronized; thus, this advantage is addressed by the robust LiDAR-inertial odometry framework FAST-LIO. It improves unmanned ground vehicles (UGVs) and unmanned aerial

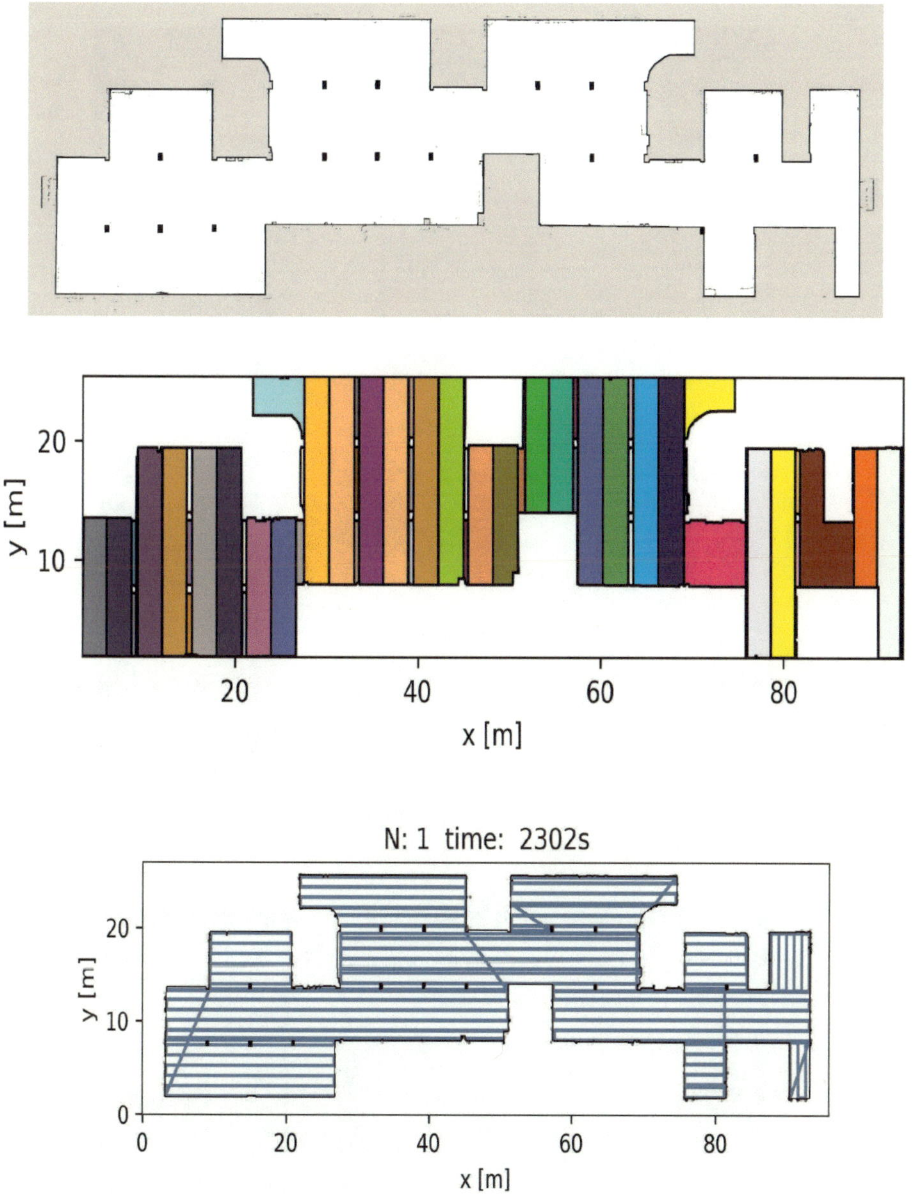

Figure 9.
Top: a grid representing obstacles in an underground garage. Pixel size is 3 cm × 3 cm. Middle: the result of a boustrophedon decomposition. Bottom: A trajectory for a single robot.

vehicles (UAVs) in 3D mapping scenarios. Our study incorporates this robust LiDAR-inertial odometry framework FAST-LIO as the front end of SLAM. The main focus is a lightweight back-end implementation of pose graph simultaneous localization and mapping (SLAM). This lightweight implementation is an alternative solution to state-of-the-art g2o or GTSAM implementations. We also elaborate iterative closest point, normal distributions transform, and their extension for multiview 3D data registration/refinement. It is based on C++ using Eigen library. This chapter also discusses

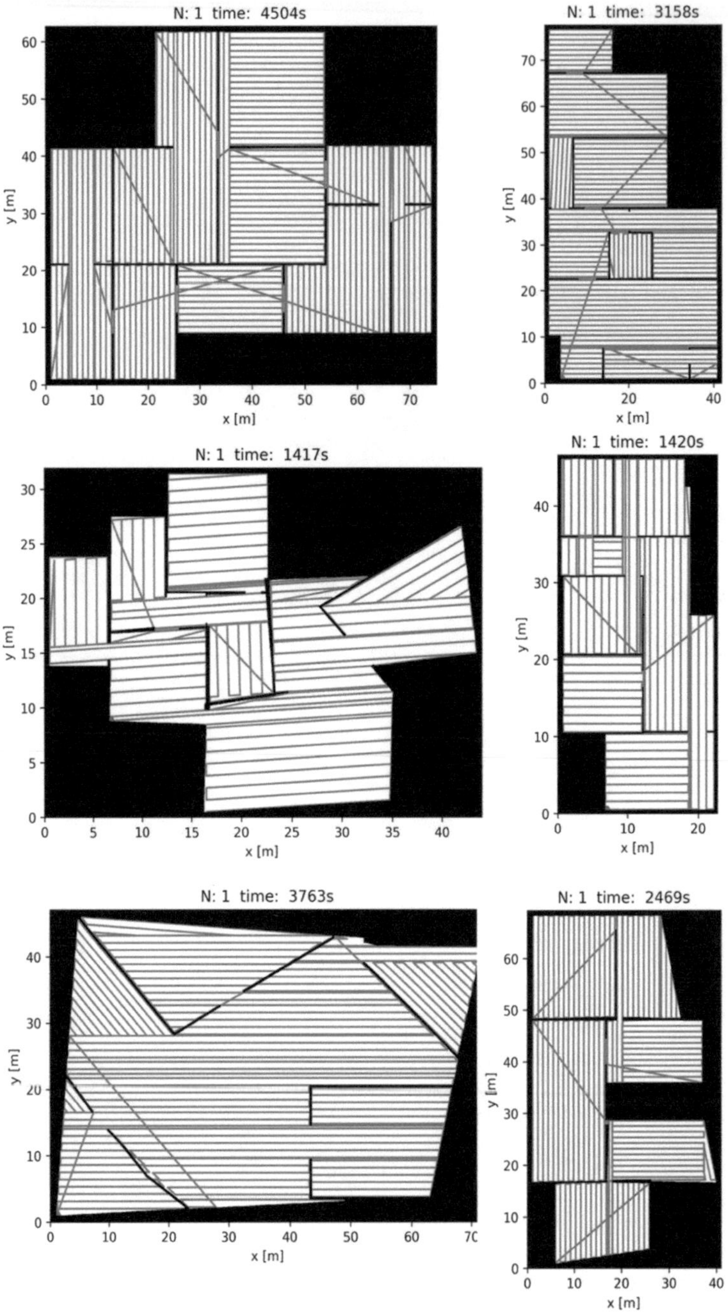

Figure 10.
Covering path plans for a small subset of the realistic, synthetic dataset of building layouts [39]. The covering area is a circle with diameter equal to 0.5 m. Notice various scales for the x and y axes and—Consequently—Various times required for the complete coverage (indicated above the plots).

path planning in already mapped environment. All software components are available as an open-source project. This chapter provides insights for useful software components for building autonomous mobile mapping robots.

Acknowledgements

The authors acknowledge the financial support of National Centre for Research and Development, project POIR.01.01.01-00-0206/17"Designing an autonomous platform which operates in an industrial production environment."

Author details

Janusz Będkowski* and Jacek Szklarski
Institute of Fundamental Technological Research, Polish Academy of Sciences, Poland

*Address all correspondence to: januszbedkowski@gmail.com

References

[1] Janusz Bedkowski. Hdmapping. 2022. Available from: https://github.com/MapsHD/HDMapping

[2] Janusz Bedkowski. Observation equations. 2022. Available from: github.com/JanuszBedkowski

[3] Wei Xu and Fu Zhang. Fast-Lio: A Fast, Robust Lidar-Inertial Odometry Package by Tightly-Coupled Iterated Kalman Filter, 2020

[4] Yang L, Fan J, Liu Y, Li E, Peng J, Liang Z. A review on state-of-the-art power line inspection techniques. IEEE Transactions on Instrumentation and Measurement. 2020;**69**(12):9350-9365

[5] Hercik R, Byrtus R, Jaros R, Koziorek J. Implementation of autonomous mobile robot in smartfactory. Applied Sciences. 2022;**12**(17):8912

[6] Nagatani K, Endo D, Watanabe A, Koyanagi E. Design and development of explosion-proof tracked vehicle for inspection of offshore oil plant. In: Hutter M, Siegwart R, editors. Field and Service Robotics, Results of the 11th International Conference, FSR 2017, Zurich, Switzerland, 12–15 September 2017. Vol. volume 5 of Springer Proceedings in Advanced Robotics. Springer; 2017. pp. 531-544

[7] Zhang Zhonglin F, Bin LL, Encheng Y. Design and function realization of nuclear power inspection robot system. Robotica. 2021;**39**(1):165-180

[8] Nagatani K, Kiribayashi S, Okada Y, Otake K, Yoshida K, Tadokoro S, et al. Emergency response to the nuclear accident at the Fukushima daiichi nuclear power plants using mobile rescue robots. Journal of Field Robotics. 2013;**30**(1):44-63

[9] Horowitz MC, Kahn L, Macdonald J, Schneider J. Covid-19 and public support for autonomous technologies—Did the pandemic catalyze a world of robots? PLoS One. 2022;**17**(9):1-18

[10] Lin J, Zhang F. R^3 live: A robust, real-time, rgb-colored, lidar-inertial-visual tightly-coupled state estimation and mapping package. In: 2022 International Conference on Robotics and Automation, ICRA 2022, Philadelphia, PA, USA, May 23–27, 2022. IEEE; 2022. pp. 10672-10678

[11] Wang Y, Lou Y, Zhang Y, Song W, Huang F, Zhiyong T. A robust framework for simultaneous localization and mapping with multiple non-repetitive scanning lidars. Remote Sensing. 2021;**13**(10):2015

[12] Li K, Li M, Hanebeck UD. Towards high-performance solid-state-lidar-inertial odometry and mapping. IEEE Robotics and Automation Letters. 2021; **6**(3):5167-5174

[13] Kelly C, Wilkinson B, Abd-Elrahman A, Cordero O, Andrew Lassiter H. Accuracy assessment of low-cost lidar scanners: An analysis of the velodyne hdl32e and livox mid40 temporal stability. Remote Sensing. 2022;**14**(17):4220

[14] Thrun S. Simultaneous Localization and Mapping. Berlin Heidelberg, Berlin, Heidelberg: Springer; 2008. pp. 13-41

[15] Kümmerle R, Grisetti G, Strasdat H, Konolige K, Burgard W. G2o: A general framework for graph optimization. In: ICRA. IEEE; 2011. pp. 3607-3613

[16] Michael Kaess. Gtsam library, 2015

[17] Besl PJ, McKay ND. A method for registration of 3-d shapes. IEEE

Transactions on Pattern Analysis and Machine Intelligence. 1992;**14**(2): 239-256

[18] Biber P, Strasser W. The normal distributions transform: A new approach to laser scan matching. In: Proceedings 2003 IEEE/RSJ International Conference on Intelligent Robots and Systems (IROS 2003) (Cat. No.03CH37453). Vol. 3. 2003. pp. 2743-2748

[19] Jihua Zhu, Di Wang, Jiaxi Mu, Huimin Lu, Zhiqiang Tian, and Zhongyu Li. 3dmndt:3d Multi-View Registration Method Based on the Normal Distributions Transform, 2021

[20] Bosse M, Zlot R. Continuous 3d scan-matching with a spinning 2d laser. In: ICRA. IEEE; 2009. pp. 4312-4319

[21] Kaul L, Zlot R, Bosse M. Continuous-time three-dimensional mapping for micro aerial vehicles with a passively actuated rotating laser scanner. Journal of Field Robotics. 2016;**33**(1): 103-132

[22] Lin H-Y, Huang Y-C. Collaborative complete coverage and path planning for multi-robot exploration. Sensors. 2021; **21**(11):3709

[23] Iqbal J, Tahir AM, Islam R u, Nabi R u. Robotics for nuclear power plants — Challenges and future perspectives. In: 2012 2nd International Conference on Applied Robotics for the Power Industry (CARPI). 2012. pp. 151-156

[24] Woohyeon Moon, Bumgeun Park, Sarvar Hussain Nengroo, Taeyoung Kim, and Dongsoo Har. Path planning of cleaning robot with reinforcement learning, 2022 IEEE International Symposium on Robotic and Sensors Environments (ROSE), Abu Dhabi, United Arab Emirates, IEEE, 2022

[25] Yamauchi B. A frontier-based approach for autonomous exploration. In: Proceedings 1997 IEEE International Symposium on Computational Intelligence in Robotics and Automation CIRA'97.'Towards New Computational Principles for Robotics and Automation'. IEEE; 1997. pp. 146-151

[26] Belavadi SS, Beri R, Malik V. Frontier exploration technique for 3d autonomous slam using k-means based divisive clustering. In: 2017 Asia Modelling Symposium (AMS). IEEE; 2017. pp. 95-100

[27] Almadhoun R, Taha T, Seneviratne L, Zweiri Y. A survey on multi-robot coverage path planning for model reconstruction and mapping. SN Applied Sciences. 2019;**1**(8):1-24

[28] Yan Z, Jouandeau N, Cherif AA. A survey and analysis of multi-robot coordination. International Journal of Advanced Robotic Systems. 2013;**10**

[29] Arkin EM, Fekete SP, Mitchell JSB. Approximation algorithms for lawn mowing and milling. Computational Geometry. 2000;**17**(1–2):25-50

[30] Choset H. Coverage for robotics–a survey of recent results. Annals of Mathematics and Artificial Intelligence. 2001;**31**(1):113-126

[31] Galceran E, Carreras M. A survey on coverage path planning for robotics. Robotics and Autonomous Systems. 2013;**61**(12):1258-1276

[32] Saeedi S, Trentini M, Seto M, Li H. Multiple-robot simultaneous localization and mapping: A review. Journal of Field Robotics. 2016;**33**(1):3-46

[33] Meurer A, Smith CP, Paprocki M, Čertík O, Kirpichev SB, Rocklin M, et al.

Sympy: Symbolic computing in python. PeerJ Computer Science. 2017;**3**:e103

[34] Marquardt DW. An algorithm for least-squares estimation of nonlinear parameters. SIAM Journal on Applied Mathematics. 1963;**11**(2):431-441

[35] Szklarski J. Multi-robot coverage with reeb graph clustering and optimized sweeping patterns. Computer Assisted Methods In Engineering And Science. 2022;**29**(4):379-395

[36] Choset H. Coverage of known spaces: The boustrophedon cellular decomposition. Autonomous Robots. 2000;**9**(3):247-253

[37] Nielsen LD, Sung I, Nielsen P. Convex decomposition for a coverage path planning for autonomous vehicles: Interior extension of edges. Sensors. 2019;**19**(19):4165

[38] Bähnemann R, Lawrance N, Chung JJ, Pantic M, Siegwart R, Nieto J. Revisiting boustrophedon coverage path planning as a generalized traveling salesman problem. In: Field and Service Robotics. Springer; 2021. pp. 277-290

[39] Li T, Ho D, Li C, Zhu D, Wang C, Meng MQ-H. Houseexpo: A large-scale 2d indoor layout dataset for learning-based algorithms on mobile robots. In: In 2020 IEEE/RSJ International Conference on Intelligent Robots and Systems (IROS). IEEE; 2020. pp. 5839-5846

Chapter 4

Coverage Technology of Autonomous Mobile Mapping Robots

SeungHwan Lee

Abstract

The coverage technique is one of the essential applications of autonomous mobile mapping robots. There are various approaches for coverage depending on the model (model/non-model), robot systems (single/multi), and its purpose (patrol/cleaning). Coverage components include viewpoint generation and path planning approaches, which are described as CPP research work. Particularly, in surveillance systems, coverage techniques, such as spanning tree, cyclic coverage, and area-based coverage, are reviewed specifically, which can be expanded for multi-robot systems. In addition, required coverage techniques according to conditions for intelligent surveillance systems are summarized. Lastly, several issues on coverage, specifically cyclic coverage, are described and considered.

Keywords: coverage, surveillance systems, issues on coverage, multi-robot patrol systems

1. Introduction

Coverage path planning (CPP) is the process of calculating a viable path through all points in a region of interest to scan or investigate a region of interest in the environment [1]. As shown in **Figure 1**, it is largely applied to the coverage problem of patrol robots and cleaning robots.

Early CPP studies [2] defined six robot requirements as follows:

1. Robot should move through all the points in the target area covering it completely.

2. Robot should cover the region without overlapping paths.

3. Continuous and sequential operation without any repetition of paths is required.

4. Robots should avoid all obstacles.

5. Simple motion trajectories (e.g., straight lines or circles) should be used (for simplicity in control).

6. An "optimal" path is desired under available conditions.

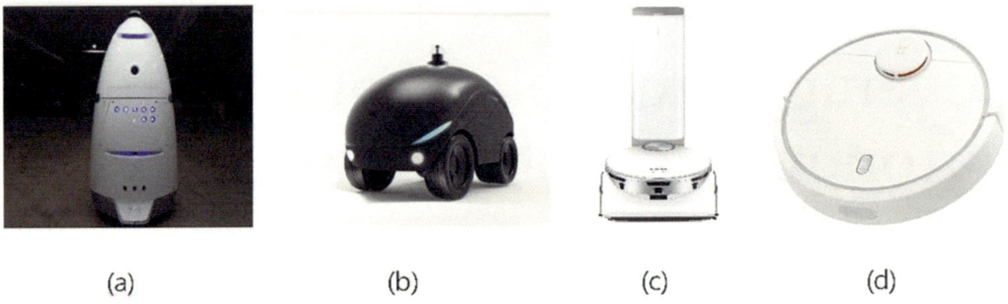

Figure 1.
Various robot platforms with coverage algorithms. (a) K5 robot, (b) D-Bot, (c) Samsung robot cleaner, and (d) Mi robot cleaner.

In addition, it requires robot positioning, autonomous driving capabilities, path planning, detection, and recognition. However, these conditions are not always possible in complex environments. Thus, sometimes only a few conditions are needed.

As aforementioned, CPP is the process of navigating or exhaustively searching a workspace. The CPP must also determine all locations to be visited while avoiding all possible obstacles [2, 3]. Essentially, applications, such as structural painting, object reconstruction, lawn mowing, surveillance, geospatial mapping, agricultural surveying, and floor cleaning, require full coverage. In general, in CPP, it is common to construct a model in real time using a sensor mounted on a robot and to provide information, such as a region of interest in advance [4–7].

The three main components of CPP are viewpoint generation, path planning, and coverage integrity quantification. In each implementation, CPP can be performed online or offline. Offline CPP requires a reference model. On the other hand, in the case of online CPP, since it is necessary to utilize sensor information, a step to process sensor information is required.

2. Model/non-model-based CPP

There are several research methods for performing CPP using a single robot. These methods can be broadly divided into model-based approaches and non-model-based approaches. In Ref. [8], a heuristic-based approach based on the art gallery problem (AGP) approach [9] was proposed. **Figure 2** shows a visual representation of the AGP. In addition, this method solved the traveling salesman problem (TSP) to generate optimized routes for each structure and then randomly assign them based on the time constraints of the UAV and the travel routes of the traveling salesman.

There is also an example of performing a TSP algorithm using PSO [10]. PSO is a particle swarm optimization technique that progressively finds an optimal solution using N particles. The solution giving the best score among particles is the output of the PSO. **Figure 3(a)** shows the structure of the PSO algorithm. It can be seen that the algorithm is configured in the form of finally finding the optimal solution while optimizing the objective function until the termination condition is satisfied. **Figure 3(b)** shows how each particle converges.

The second approach of the single-robot CPP approach follows a non-model-based approach. The work presented in Ref. [12] proposed an extended CPP approach [13] by utilizing surface information to plan coverage paths online using

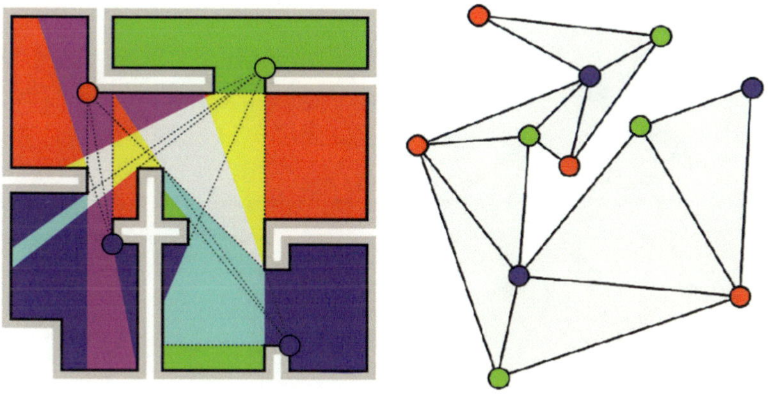

Figure 2.
Visual representation of AGP [9]. This problem refers to the problem of protecting the entire museum with the minimum number of guards. In general, AGP is to ensure that the guard represented by a point on a polygonal map is sufficient to detect all spaces.

```
Algorithm 1: Particle Swarm Optimization
Input  : Objective function f : 𝒳 → ℝ,
         Termination condition ψ : 𝒳 → 𝔹,
         Population size: N,
         Lower and upper bounds of the solution: b_lb and b_ub,
         Maximum influence values φ1 and φ2
Output: Best solution g
// Step 1: Initialization.
Randomly initialize the population 𝒫 = {x1, x2, ..., xN}.
Randomly initialize the particle's velocity within [b_lb, b_ub].
repeat
    for i ∈ {1, 2, 3, ..., N} do
        // Step 2: Velocity Calculation.
        // d is a dimensionality of the input space 𝒳.
        Generate a random vector r1 ~ U[0, φ1]^d
        Generate a random vector r2 ~ U[0, φ2]^d
        v_i^(k+1) ← v_i^(k) + r1(p_i − x_i^(k)) + r2(g − x_i^(k))
        // Step 3: Position Update
        x_i^(k+1) ← x_i^(k) + v_i^(k+1)
        // Step 4: Evaluation
        if f(x_i^(k+1)) < f(p_i) then
            p_i ← x_i
            if f(p_i) < f(g) then
                g ← p_i
            end
        end
    end
until ψ(𝒳) == TRUE;
return g
```

(a)

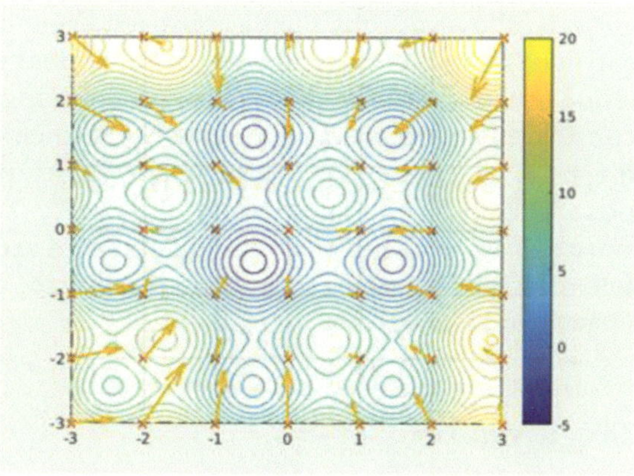

(b)

Figure 3.
The representation of the PSO algorithm and convergence [11]. The pseudo-code of PSO is represented in (a). The movements of particles are shown in (b). (a) PSO algorithm (b) visualization of PSO.

truncated signed distance fields (TSDF). The search space is divided into a surface area and a cuboid area, which are used to create a volume map of the bounding area. The volume map is used to calculate the information gain by considering the cube volume and path length. The Hamiltonian path problem is to compute the visit order for each cuboid while generating the path using the generalized TSP. As shown in **Figure 4**, the Hamiltonian path problem can be solved by finding out a path through all vertices once in graph theory.

In Ref. [15], another non-model-based approach, that is, a search algorithm, was presented that selects the next best view (NBV) that maximizes the predicted information gain, taking into account distance and battery life cost. The proposed task dynamically builds a hull surrounding a predefined bounding box that updates based on new information. The visited points are uniformly sampled with a fixed number pointing to the vertical axis through the center of the bounding box. The planning

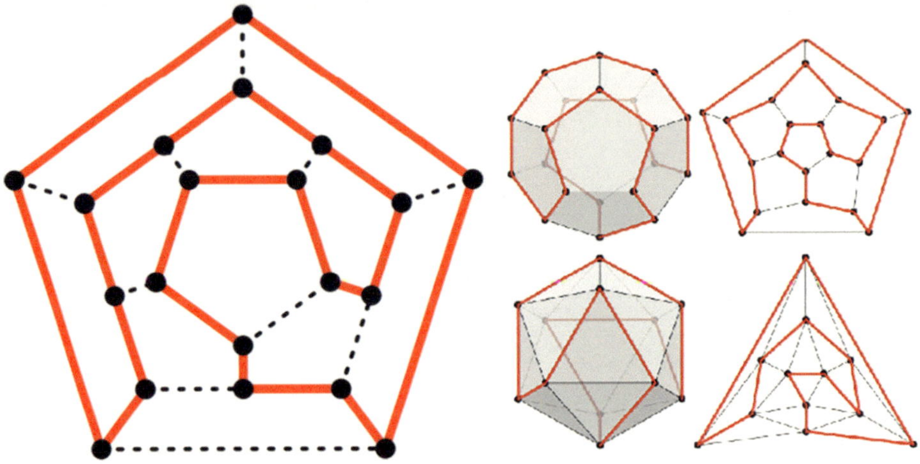

Figure 4.
Hamiltonian path generation problems [14]. Hamiltonian path generation results for various examples are shown.

approach follows a probabilistic approach using utility functions that reduce 3D reconstructed model uncertainty, divert combat paths, and generate safe paths based on time constraints. The energy aspect of the non-model-based single-robot approach is considered an important part, especially since the CPP is performed online. Using this type of approach on a single robot makes it difficult to achieve high coverage ratios and increases computational complexity. This difficulty arises from the complexity of environments that contain enclosed areas that are difficult to find with a single robot and require a lot of time to navigate. Single-robot CPP approaches were reviewed and analyzed in Ref. [16].

3. Single/multi-robot

For large areas and structures, leveraging a multi-robot CPP strategy can be a huge advantage to quickly achieve full coverage. Using one robot to cover a large structure or a large area has various disadvantages, such as time, length, robot energy, and quality and quantity of information [6, 7]. The multi-robot CPP approach follows the same approach as single-robot coverage, but requires additional factors and requirements to be considered. These factors include the type of collaboration, information sharing, robustness of agent failure handling, level of autonomy, robot durability, and task assignment. It is necessary to explore different approaches utilizing multi-robot systems in terms of CPP components, including viewpoint generation and path planning approaches. Path planning approaches can be divided into grid-based navigation approaches, geometric approaches, reward-based approaches, NBV approaches, and random incremental planning approaches.

Performing CPP using a multi-robot system for model reconstruction and mapping requires various aspects to be considered. Two key CPP-related aspects for creating feasible coverage routes include viewpoint generation and path planning. The remaining aspects are communication/task assignment and mapping in multi-robot systems, which are important to model or construct regions of interest using the collected data.

3.1 Viewpoint generation

In most studies, the coverage search method is largely divided into the model-based method and the non-model-based method. Model-based methods rely on a reference model of the environment or structure provided first, whereas non-model-based ones perform planning and exploration without prior knowledge of the structure or environment [4, 5]. Based on these classifications, viewpoints are generated to form the search space of the planner. Some methods of generating them are uniformly generated due to the existence of structural or regional models and their dependencies on specific regional or structural models. Other types of viewpoint generators are also randomly assigned due to lack of knowledge of structural or domain models. Viewpoint generation is treated as important in the multi-robot CPP process because it aims to output a set of optimized routes that represent an acceptable set of viewpoints containing the structure or environment of interest. Depending on the search method used and the scope of its application, various techniques for performing viewpoint generation are mentioned in the relevant papers.

3.2 Path planning

The multi-robot CPP approach is being pursued in various studies describing challenging problems and proposed solutions. All of these approaches have a similar goal of providing a collision-free path that achieves the full extent of a structure or area. As a representative method, the visibility graph is used for path generation [17]. Visibility graphs contain a set of points and obstacles where nodes represent locations, and edges consist of line segments that do not pass-through obstacles. Examples of visibility graphs and generated paths are shown in **Figure 5**.

Geometric approaches such as shortest Euclidean path finding and polygonal area coverage are used in many domains. The most popular geometric-based method in multi-robot CPP is the Voronoi diagram. In [19], a dynamic path planning approach was proposed for heterogeneous multi-robot sensor-based coverage (DPP MRSBC)

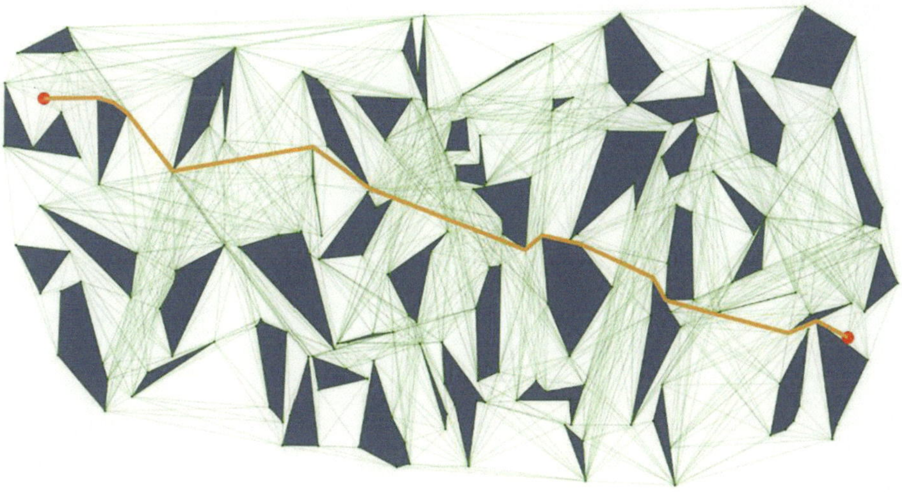

Figure 5.
Examples of visibility graphs and generated paths [18].

considering energy capacity. The environment was modeled as a generalized Voronoi diagram (GVD) that obscures the edges of the diagram. The proposed algorithm starts with an undirected graph and creates a directed subgraph.

3.3 CPP research work

The main components of the CPP process used in the work recently investigated in this article are summarized in **Figure 6**. **Figure 6** summarizes recent research on multi-robot CPP, which includes the evaluation metrics for the environment type, algorithm processing technique, viewpoint generation method, application path generation method, and application method. According to the researched

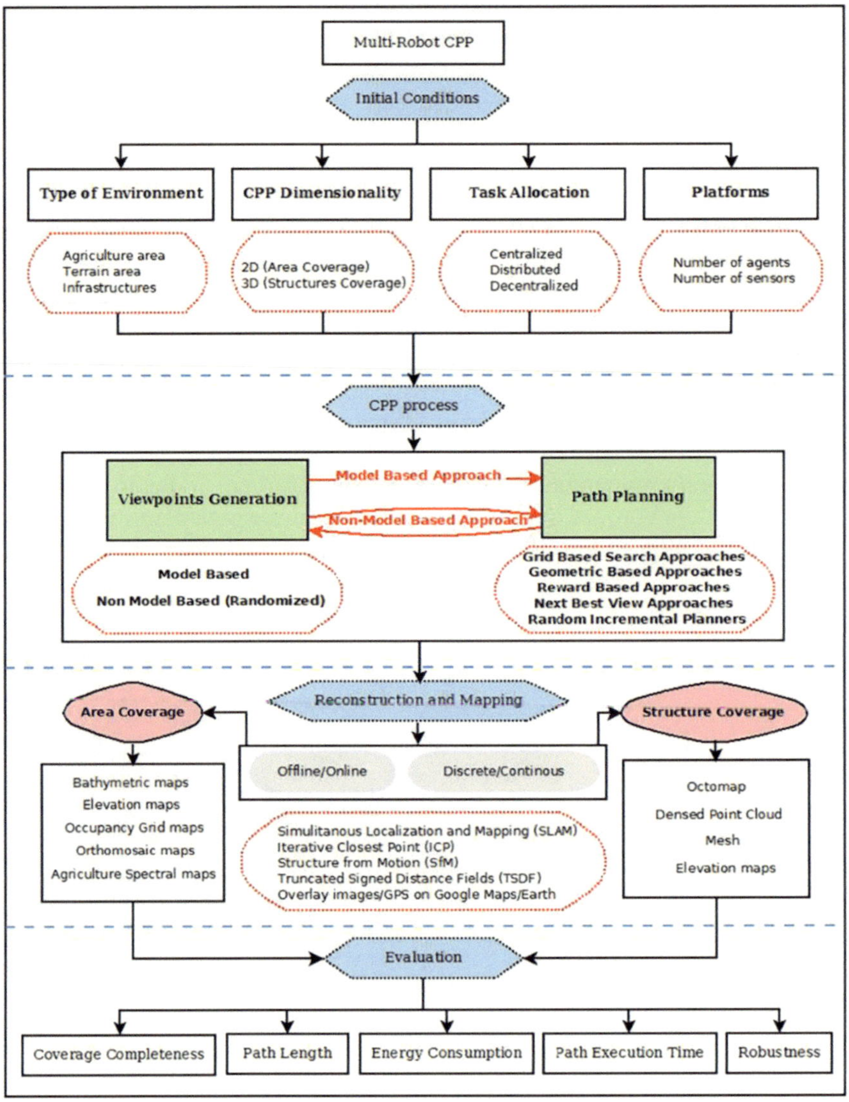

Figure 6.
Summary of CPP research work [1]. It includes the evaluation metrics for the environment type, algorithm processing technique, viewpoint generation method, application path generation method, and application method.

literature, there are many problems that block the progress of efficient multi-robot cooperative CPP. These issues include heterogeneity, prioritization, robustness, communication, adaptability, open systems, collective intelligence, and multi-robot scheduling.

4. Coverage for patrol robots

With the development of robot technology, research and interest in unmanned patrol robots have been increasing in various fields that require monitoring and security, such as social safety and national defense [20–24]. In particular, research on multi-object systems consisting of two or more patrol robots in a wide area is being actively conducted [25, 26]. These systems were essentially trying to solve the problem of multi-robot patrols. The challenge is to find the optimal solution for given tasks, that is, monitoring, information gathering, object discovery, anomaly detection, and so on (**Figure 7**).

Several studies have been attempted to solve the monitoring and security problems of multi-agent systems, and through this, the necessary parts for an intelligent monitoring system are as follows. Given an environment map, we need to (1) extract nodes automatically to generate a robot roadmap. In particular, when generating nodes, it is necessary to obtain sophisticated node extraction results by reflecting the normal vector direction of the map and the sensing range of the sensor. (2) You need to solve the TSP to create a full patrol route. (3) In order to assign a patrol route to a multi-agent system, it is necessary to represent the relation between the maximum number of robots, the maximum patrol period, and the maximum speed of the robots. As a result, the patrol paths assigned to each agent can be derived. (4) If the environment map or the obstacle probability map according to the density of obstacles other than the point of interest is given, it is necessary to change the maximum velocity of the robot in the corresponding area or to allocate a dedicated robot. A summary of this can be found in **Table 1**.

Among them, the study of P. Fazli [26] is shown in **Figure 8**. Several concepts are also represented. First, the path between robots should not overlap, and it is an approach that increases the frequency of visits by reducing the visit period. The method is done according to following process.

Figure 7.
The components of the multi-robot surveillance system [20]. (a) Multi-robot monitoring. (b) Multi-robot patrol mission.

Map condition	Robot condition	Required essential techniques
Environments	# of robots, patrol interval, patrol frequency, robot maximum vel, sensing range	Graph reconstruction for robot road map path generation for N robots
Environments + Obstacle information	# of robots, patrol interval, patrol frequency, robot maximum vel, sensing range	Obstacle-based local path regeneration
environments + Region of interests	# of robots, patrol interval, patrol frequency, robot maximum vel, sensing range	Path regeneration technology considering dedicated robot assignment

Table 1.
Coverage techniques and conditions for intelligent surveillance systems.

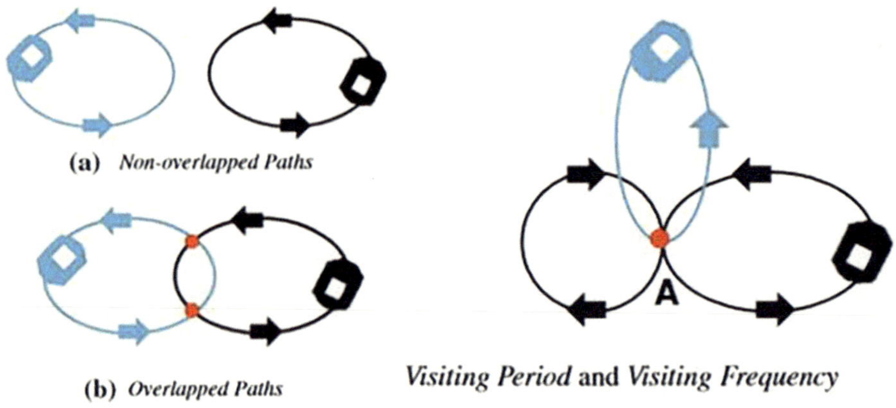

Figure 8.
Schematic diagram of the path overlap problem and the visiting frequency problem [26].

1. Select the position where the fixed guard should be located as the region of interest and use the field of view of the robot at this time.

2. Complete graph G based on the location of the fixed guards and obstacles of the given polygon. (using visibility graphs or Delaunay triangulation)—complete robotic roadmap.

3. The graph uses the reduced visibility method or the reduced CDT method to reduce the number of connected nodes.

4. Perform cluster-based coverage algorithm: Divide the given graph into small pieces of area and assign them to robots. This will soon be the full patrol route for each robot. (There are uniform clustering method, edge-based clustering method, and node-based clustering method.)

5. In the cluster-based method, patrol routes may not be made similar, so a circular coverage algorithm can be performed for equidistant route assignment. (This method creates the shortest traversal path and assigns it uniformly to each robot.)

6. Finally, the path generation algorithm is performed using a double-minimum spanning tree or a linked Lin-Kernighan algorithm to create a path that traverses the start and end of the graph assigned to each robot.

4.1 Spanning tree-based coverage (STC)

The input to the STC algorithm is a constrained planar environment, partially filled with smooth and stationary obstacles. The algorithm first subdivides the working area into 2D-sized cells and discards cells partially covered by obstacles. The graph structure G(V, E) is defined as a line segment that defines the center point of each cell as node V and the centers of adjacent cells with edge E. The following algorithm constructs a spanning tree for G and uses this tree to generate coverage paths as shown in **Figure 9**.

4.2 Cyclic coverage algorithm

The cyclic coverage suggested by P. Fazli is similar to the cluster-based coverage algorithm, and the cyclic coverage approach finds the guards, builds a graph (VG), and then reduces the graph, named reduced Vis. It uses the Chained LinKernighan algorithm to generate a path for the entire reduced Vis. The proposed algorithm then distributes the robot equidistantly around the tour and moves it recursively. The cyclic coverage approach produces an optimal or near-optimal solution for a single robot in terms of full path length and total worst-case visit duration.

4.2.1 Issues of cyclic coverage

There are about seven issues that are mainly dealt with in the cyclic coverage problem as shown in **Table 2**. The goal can be reviewed as a problem of minimizing the worst-case frequency or maximizing a random patrol target. However, most coverage problems were handled by minimizing the worst visiting frequencies. The environment can be largely divided into an indoor environment (corridor space) and an outdoor environment (open space). The third is whether there are any restrictions on the sensor. In practice, it is necessary to approach the problem of limited detection range and deal with it in detail. Agents can be viewed in static or dynamic environments and

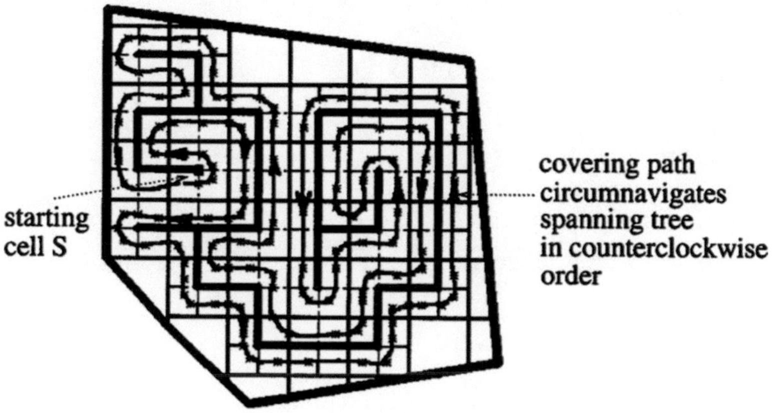

Figure 9.
Coverage path generation result [27]. It is based on the spanning tree for graph G.

Interests	Considerations
Objective	a. Minimize the worst frequency b. Maximize the probability of detection
Environment	a. Open space b. Corridor space (complex)
Sensors	a. Limited range b. Unlimited range (like vision)
Agents	a. Goal oriented b. Reactive
Controller	a. Centralized control b. Distributed control
Patrol policy	a. Area patrolling b. Boundary patrolling
Solution	a. Optimal b. Near-optimal

Table 2.
Issues of cyclic coverage methods.

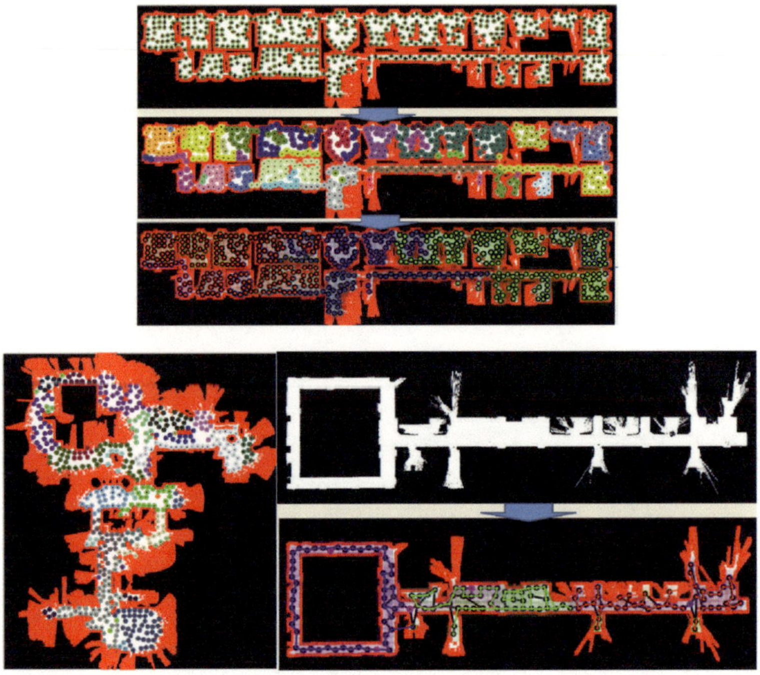

Figure 10.
The simulation results of the cyclic coverage method. The SLAM maps are built using the SLAM dataset.

are usually defined in a goal-oriented fashion, which can be different depending on the given problem. If there are no communication restrictions, centralized control can be conducted to the robots, and the patrol policy can be defined as an area patrol that covers all areas. If it is a border patrol problem, then the method produces how agents approach the border. Since in most cases the exact solution is not known, the goal is to get near-optimal in terms of time and distance.

4.2.2 Simulation result of cyclic coverage

The cyclic coverage method [28] can be applied to various maps as shown in **Figure 10**. (SLAM data set published on Ref. [29]). Three robots were considered for three maps in this simulation. The first figure in **Figure 10** is the result of generating a graph (consisting of nodes marked in green and edges) considering the detection range of the robot. By creating a single robot path from this graph and removing redundant paths (leaving unique paths), the path can be represented piecewise. The final result is obtained by merging a number of paths equal to the number of robots and assigning those paths to each robot. From the simulation results in **Figure 10**, it can be seen that each robot's coverage area is properly allocated.

5. Area-based coverage

Path distribution techniques can give poor results depending on the presence and geographic location of overlapping paths. To solve this problem, the process of area allocation can be applied. One method is the DARP (region segmentation based on the robot's initial position) method [30]. This method divides the area based on Voronoi segmentation and the initial position of the robot. It also scales the area through the metric function. The method has been improved according to the form of versions, such as 0.5, 1.0, and so on.

6. Conclusion

In this chapter, coverage techniques have been reviewed in terms of the model, robot systems, and their purpose by showing their procedures and simulation results. Particularly, in surveillance systems, coverage techniques, such as spanning tree, cyclic coverage, and area-based coverage, were described specifically, which can be expanded for multi-robot patrol systems. In addition, several issues on coverage were described and considered.

Acknowledgements

This research was a part of the project titled "Research on Co-Operative Mobile Robot System Technology for Polar Region Development and Exploration," funded by the Korean Ministry of Trade, Industry and Energy (1525011633).

Author details

SeungHwan Lee
School of Electronic Engineering, Kumoh National Institute of Technology, Gumi, Republic of Korea

*Address all correspondence to: leesh@kumoh.ac.kr

References

[1] Almadhoun R, Taha T, Seneviratne L, Zweiri Y. A survey on multi-robot coverage path planning for model reconstruction and mapping. SN Applied Sciences. 2019;**1**(847):1-24

[2] Galceran E, Carreras M. A survey on coverage path planning for robotics. Robotics and Autonomous Systems. 2013; **61**:1258-1276

[3] Valente J, Barrientos A, Cerro JD. Coverage path planning to survey large outdoor areas with aerial robots: A comprehensive analysis, In: Introduction to modern robotics II. iConcept Press. 2011. ISBN: 978-1-463789-442

[4] Scott WR. Model-based view planning. Machine Vision and Applications. 2009;**20**:47-69

[5] Scott WR, Roth G, Rivest JF. View planning for automated three-dimensional object reconstruction and inspection. ACM Computing Surveys. 2003;**35**(1):64-96

[6] Choset H. Coverage for robotics—A survey of recent results. Annals of Mathematics and Artificial Intelligence. 2001;**31**(1-4):113-126

[7] Muddu RSD, Wu D, Wu L. A frontier based multirobot approach for coverage of unknown environments. In: 2015 IEEE international conference on robotics and biomimetics, IEEE-ROBIO 2015. 2015. pp. 72-77

[8] Doitsidis L, Weiss S, Renzagli A, Achtelik MW, Kosmatopoulos E, Siegwart R, et al. Optimal surveillance coverage for teams of micro aerial vehicles in GPS-denied environments using onboard vision. Auton Robots. 2012;**33**(1-2):173-188

[9] ART Gallery Problem. Available from: https://en.wikipedia.org/wiki/Art_gallery_problem

[10] Phung MD, Quach CH, Dinh TH, Ha Q. Enhanced discrete particle swarm optimization path planning for UAV vision-based surface inspection. Automation in Construction. 2017;**81**:25-33

[11] PSO algorithm. Available from: https://en.wikipedia.org/wiki/Particle_swarm_optimization

[12] Song S, Jo S. Surface-based exploration for autonomous 3D modeling, In: IEEE International Conference on Robotics and Automation (ICRA'18), Brisbane, Australia. 2018. pp. 4319-4326

[13] Song S, Jo S. Online inspection path planning for autonomous 3D modeling using a micro-aerial vehicle. In: Proceedings—IEEE International Conference on Robotics and Automation. 2017. pp. 6217-6224

[14] Hamiltonian Path. Available from: https://en.wikipedia.org/wiki/Hamiltonian_path

[15] Palazzolo E, Stachniss C. Effective exploration for MAVs based on the expected information gain. Drones. 2018;**2**(1):9

[16] Almadhoun R, Taha T, Cai G. A survey on inspecting structures using robotic systems. International Journal of Advanced Robotic Systems. 2016;**13**(6):1-18

[17] Visibility Graph. Available from: https://en.wikipedia.org/wiki/Visibility_graph

[18] Shortest Path Planning on Visibility Graph. Available from: https://fribbels.github.io/shortestpath/writeup.html

[19] Zheng X, Koenig S, Kempe D, Jain S. Multirobot forest coverage for weighted and unweighted terrain. IEEE Transactions on Robotics. 2010;**26**(6):1018-1031

[20] Yan C, Zhang T. Multi-robot patrol: A distributed algorithm based on expected idleness. International Journal of Advanced Robotic Systems. 2016;**13**(6):1-12

[21] Witwicki S, Castillo JC, Messias J, Capitan J, Melo F, Limá PU, et al. Autonomous surveillance robots a decision-making framework for networked multi agent systems. IEEE Robotics & Automation Magazine. 2017; **24**(3):52-64

[22] Shermer TC. Recent results in art galleries[geometry]. Proceedings of the IEEE. 1992;**80**(9):1384-1399

[23] Nilsson BJ. Guarding art galleries-methods for mobile guards. [PhD thesis]. Lund University. 1995

[24] Ye Y, Tsotsos JK. Sensor planning in 3D object search. Computer Vision and Image Understanding. 1999;**73**(2):145168

[25] Fazli P, Davoodi A, Mackworth AK. Multi-robot repeated area coverage: Performance optimization under various visual ranges. Computer and Robot Vision. 2012:298-305

[26] Fazli P, Davoodi A, Mackworth AK. Multi-robot repeated area coverage. Autonomous Robots. 2013;**34**(4):251-276

[27] Gabriely Y, Rimon E. Spanning-tree based coverage of continuous areas by a mobile robot. Annals of Mathematics and Artificial Intelligence. 2001;**31**:77-98

[28] Noh D, Choi J, Choi J, Byun D, Youngjae K, Kim H-R, et al. MASS: Multi-agent scheduling system for intelligent surveillance, UR. 2022

[29] Open SLAM. Available from: www.openslam.org

[30] DARP. Available from: https://medium.com/@athanasios.kapoutsis/darp-divide-areas-algorithm-for-optimal-multi-robot-coverage-path-planning-2fed77b990a3

Section 3

Large Scale Mapping

Chapter 5

Multi-Robot Mapping Based on 3D Maps Integration

Michał Drwiega and Elżbieta Roszkowska

Abstract

An unknown environment could be mapped more efficiently by a group of robots than a single robot. The time reduction due to parallelization is crucial in complex area mapping. There are two general solutions used in the multi-robot mapping. In the first one, robots exchange raw data from sensors. The second approach assumes that each robot creates a local map independently that is exchanged with other robots and integrated. In this chapter, we present a 3D maps integration algorithm that utilizes overlapping regions in the feature-based alignment process. The algorithm does not need any initial guess about the transformation between local maps. However, for successful integration, maps need to have a common area. We showed that the implemented method is effective in various environments. The approach has been verified in experiments with wheeled mobile robots and using public datasets with octree-based maps.

Keywords: multi-robot mapping, MR-SLAM, feature-matching, octomaps, ICP

1. Introduction

The development of autonomous mobile robots is accelerated by various applications like underground exploration [1, 2], planetary exploration, autonomous cars, search and rescue, reconnaissance, home cleaning, lawn mowing, or industrial applications. In many of these implementations, Multi-Robot Systems (MRS) have several advantages over a single robot. First of all, task execution time may be reduced due to parallelization. Moreover, the multi-robot system can provide a higher level of reliability. Even in the case of a single robot failure, other robots can complete the task. These features are crucial, for example, in search and rescue applications when robots find survivors in the extreme underground environment after accidents.

The multi-robot mapping of unknown environments can also be performed more efficiently than a single robot mapping. It is essential in the case of large and complex area mapping. Moreover, the robots can be equipped with different kinds of sensors to create more accurate world models. Nonetheless, several challenges specific to multi-robot systems arise, for instance, proper coordination of robots or handling the communication between them.

In general, there are two solutions used in multi-robot mapping. In the first one, robots exchange raw data from sensors. The second approach assumes that each robot creates a local map in the local coordinate system independently. Then local maps are exchanged between robots and integrated into one global map [3]. This chapter focuses on the second solution.

The key part of the map merging process is an estimation of the transformations between maps. The map alignment process is more challenging than consequent sensor data frames matching because of larger displacement between maps or measurements. There are several possible solutions on how to find this transformation. One of them assumes using the robots' initial poses and their local localization systems. However, it is not possible in all cases because of localization drift.

Another approach is to exchange and merge maps only during robot meetings. It assumes the use of additional sensors or methods to detect other robots when they see each other. Even partial information could be helpful during the transformation estimation, so in some systems, only the distance between robots is calculated based on the signal time-of-flight.

It is also possible to use a feature matching-based approach to get the relative poses of robots. In this case, the features are extracted from maps and matched. However, the transformation could be estimated successfully only if maps have an overlapping area.

This paper presents the design and implementation of the global 3D maps integration method with the alignment based on feature matching which does not require an initial transformation estimation.

1.1 Related work

One of the basic tasks of multi-robot mapping is a merging of local maps from individual robots into one global map [4–6]. As mentioned before, the most challenging part of the maps merging is finding the transformation between partial maps.

Some approaches use additional information about transformation between maps. For instance, it can be acquired by visual or range measurements during the robots meeting. In ref. [7], an approach has been presented that is based on the direct measurements between robots. Each direct detection creates a hypothesis that is validated during the next meetings of the same robots but in other locations. Maps are merged only if robots meet again and the hypothesis is accepted.

Map merging methods are also based on the idea of finding and matching the overlapping area in the maps. Such overlapping areas are not known before. In ref. [8], the system for detection of overlapping regions in maps created with ceiling-vision-based SLAM has been described. The algorithm can detect the overlapping regions and estimate transformations between partial maps.

Another approach has been included in ref. [9]. It uses an omnidirectional visual system and the initial version was intended for only one robot that creates partial maps. Nevertheless, the method can be applied also to multi-robot systems. The algorithm uses a vision system to generate coarse transformations between partial maps. The additional level of validation is based on the calculated bounding boxes with the Haar-based place recognition method.

The next group includes methods based on sensor measurements matching, for example, scan matching. In [10], the 2D local maps were merged with SIFT (*Scale-Invariant Feature Transform*) algorithm that allows to extract, describe and match features. Another optimization technique that was used is the ICP (*Iterative Closes Point*) [11, 12] which finds a rigid transformation between two point sets. It has been utilized to find a transformation between consecutive 2D scans during the map creation process.

In ref. [13], it has been presented as a spectral information-based method dedicated to 2D maps. It assumes that the map merging problem is a binary image matching problem. The algorithm allows using for instance Hough transforms to decompose the transformation into separate rotation and translation. Another method that uses geometric and topological similarities of vertices and edges to find a match between two maps has been presented in ref. [14].

A major part of the mentioned methods was supposed to be used with 2D maps. However, 3D maps have received much attention in recent years due to the growing demand for robotics services in complex environments to coexist with humans but also challenging extreme underground or underwater scenarios. The world representation in three dimensions allows robots to operate in multi-level buildings, in cluttered rooms but also in rough terrain. Moreover, such maps have better performance in heterogeneous multi-robot systems [15–17], especially when robots are equipped with different sensors. Several studies, for example [18, 19], have proposed solutions for the octomaps [20] integration problem with local alignment methods like ICP. It is worth mentioning that the octomap [21] is a tree-based representation built upon a multiple dividing of the world into eight cubic parts. Such representation is memory efficient and could be successfully used for large environments.

The mentioned map integration methods have a significant drawback. They do not optimize solutions in the background. It means that once maps are aligned the transformation is not corrected anymore. To solve this issue graph-based methods were developed [22] that benefit from backend graph optimization.

In ref. [23], local alignment method NDT (*Normal Distributions Transform*) has been presented. The comparison of NDT with ICP [24] shows that it is more efficient and in most cases, it converges from a broader range of initial poses. However, authors have noticed also that the NDT is less predictable than ICP in cases with smaller initial pose errors.

The ref. [25] presents the 3D point sets alignment algorithm that utilizes the transformation into the Radon/Hough domain.

As in opposition to the methods based on the local features description and matching, in ref. [26], an approach with higher-level descriptors based on lines or planes has been presented. Authors have noticed that higher-level descriptors improve performance in the case of small maps overlapping.

1.2 Contribution

This chapter presents a developed global 3D maps integration algorithm with the alignment based on feature matching. In contrast to many other approaches, it does not require an initial transformation estimation and is not sensitive to the local minima. The approach is based on a classic computer vision pipeline that has been modified and applied to the 3D maps integration. Moreover, the introduced model division into submodels procedure improves the feature matching process performance and increases the efficiency of data processing in many cases.

The method uses octree-based maps (octomaps) which allow to utilize the additional information like nodes' occupancy probability in the alignment process. The approach was verified in multiple test cases based on data from real robots. Furthermore, the algorithm has been implemented in C++ and released as open-source software (*3d_map_server* [27]).

1.3 Problem statement

Let us consider a system of N robots, in which each robot creates its partial map M_{n} in a local coordinate system T_n. Each map consists of a set of N_n nodes $M_n = \{m_1, m_2, \dots, m_{N_n}\}$. The maps integration problem can be defined as the creation of the consistent model of the world M based on a set of k separate models $M' = \{M_1, \ldots M_k\}$ (**Figure 1**).

In the next part of the chapter, the problem has been narrowed down to only two input models. However, in the case of more than two maps, they can be integrated sequentially.

2. Maps integration method

The presented maps integration algorithm consists of a few processing steps (**Figure 2**). On the input, there are two 3D maps (octomaps). To integrate them successfully, they must have an overlapping area. The map merging process consists of two major steps: finding the transformation between maps and data aggregation. On the output of the algorithm, there is an integrated map.

The most complex part of the algorithm is the process of the transformation finding. It consists of three steps: model extraction, global alignment, and local alignment. In the presented pipeline, map 2 is used entirely but map 1 is used to extract n models.

In the global alignment, there are common operations for both maps, like filtration, keypoints detection, and feature description. The feature descriptors from both maps are matched to each other with dedicated algorithms. Because one map has been divided into multiple models, the result of the initial alignment consists of n hypotheses $H = \{h_1, \ldots, h_n\}$ (each for a separate model).

Each hypothesis is validated with quality measures, for example, the fitness score. Finally, the best solution is selected from the set of accepted hypotheses H_A. If at least one hypothesis is accepted then the maps merging process is continued and the final result is the transformation that transforms one of the maps into the coordinate system of the other map. Otherwise, the processing is stopped at this point.

If the solution has been found in the previous step, then it is corrected with a local alignment algorithm. To generate the final map that consists of partial maps, it is necessary to transform maps to the common coordinate system and aggregate data from them into one model.

2.1 Models extraction

As mentioned before, one map is divided into multiple rectangular blocks that are used as models. Several cases of relations between maps have to be considered (**Figure 3**). The map integration can be finished successfully only in the first three cases when the area that corresponds to the model could be found in another map. In

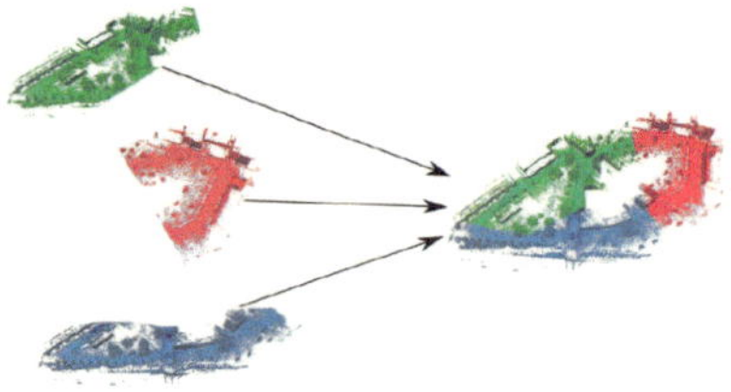

Figure 1.
Partial maps created by robots and merged into one model.

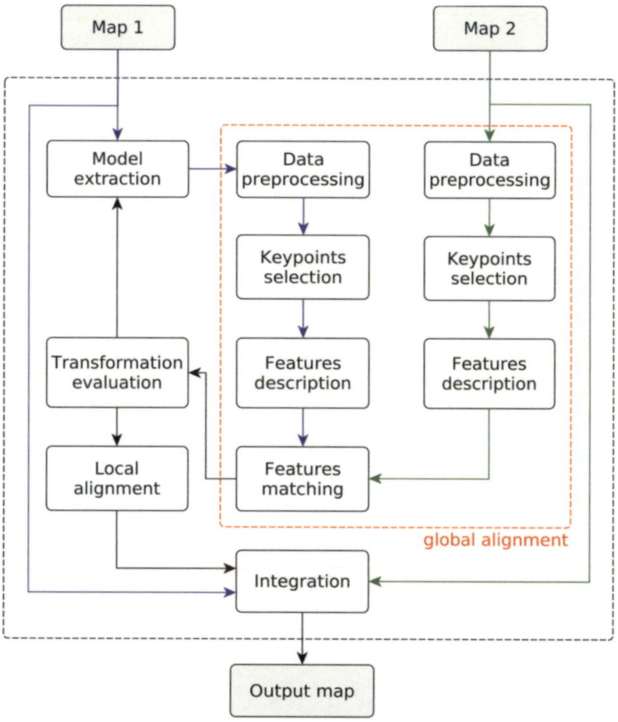

Figure 2.
The architecture of the developed maps integration algorithm.

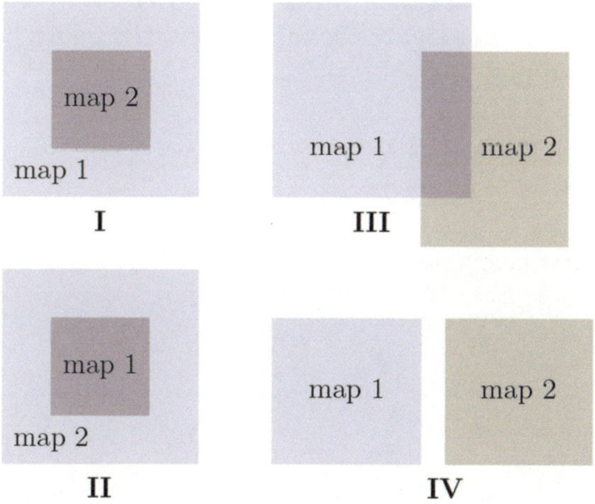

Figure 3.
Maps overlapping cases (I-III) and the case when there is no common part between maps (IV).

the last case, the integration method will be stopped because there is no common area that could be matched.

The maps integration process can be speeded up when robots start exploration from the same place and explore separate parts of the environment. Therefore, excluding the kidnapped robot problem, and a case when one of the maps is entirely included in the second one, the overlapping area begins on the borders of both maps.

Because of that, the processing of the map starts from the outside part and proceeds in a spiral toward the center of it. Concurrent models are processed in parallel until an acceptable solution is found.

If robots move on a flat surface, maps are limited on the Z axis and wider in x and y axes. In such a case, the map division in two dimensions is sufficient. Nonetheless, in specific cases like multi-level building maps, the model can be extracted from one of the maps based on the 3D grid (**Figure 4**).

2.2 Data preprocessing

The first step of data processing is pass-through filtration which rejects points that are not inside the specific area. For instance, the maps are reduced in the z-axis to get rid of the ground and reduce the number of points, and speed up further calculations.

Then, the point cloud is downsampled with a voxel grid filter. The idea behind it is to divide space into voxels with specified sizes and represent each voxel by a center point calculated as an approximation of all original points from this voxel.

The last step is outliers removal based on a statistical analysis of neighboring points. The points that do not meet the requirements are removed from the set.

2.3 Keypoints selection

The feature description is a computationally expensive operation. Therefore, the descriptors are computed only in carefully selected points - key points. The keypoints should be distinctive and repeatable to deal with noises and different points of view.

In this work, the ISS (*Intrinsic Shape Signatures*) [28] detector has been used. The ISS detector determines the relevance of the point based on eigenvalues of the matrix created for the support (local neighborhood) of a specific point. Let $X = \{x_1, x_2, \ldots, x_N\}$ be a support of the keypoint k. Then,

$$\Sigma_k = \frac{1}{N}\sum_{i=1}^{N}(x_i - p_k)(x_i - p_k)^T, \tag{1}$$

be covariance matrix calculated for the support of k, where

$$p_k = \frac{1}{N}\sum_{i=1}^{N} x_i \tag{2}$$

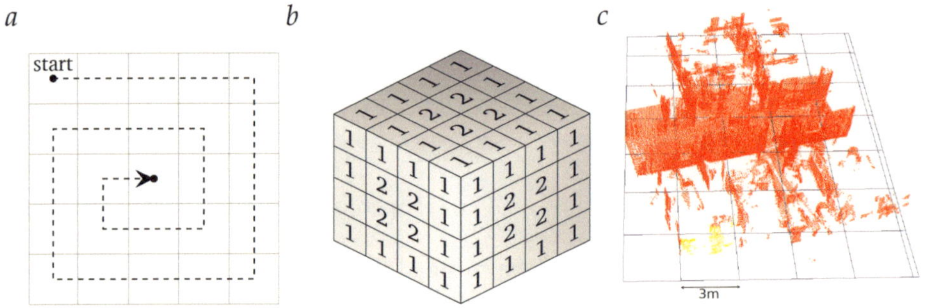

Figure 4.
The models extraction procedure with processing order - 2D (a), 3D case (b), and an example with real map (c).

is a mean point. The keypoint is relevant only if

$$\frac{\lambda_2}{\lambda_1} \leq \gamma_{12} \wedge \frac{\lambda_3}{\lambda_2} \leq \gamma_{23}, \tag{3}$$

where γ_{12} i γ_{23} are arbitrary selected parameters and $\lambda_1 < \lambda_2 < \lambda_3$ are eigenvalues of Σ_k.

2.4 Features description

Features description creates a local surface representation that could be easily compared what is crucial in the process of similar features finding. Therefore, the local descriptor is computed in each keypoint (**Figure 5**). Local descriptors describe a local neighborhood of a query point.

A local descriptor that was used is a SHOT (*Signature of Histograms of Orientations*) descriptor [29]. It uses the spherical support that is divided into spatial segments (**Figure 6**). For each segment, it is calculated a histogram representing the distribution of the $\cos\theta_i$, where θ_i is the angle between the surface normal in each point from the support and the surface normal vector n in the query point. Finally, all local histograms are combined into a vector that is the descriptor. It is also possible to utilize texture information like a point color received from the RGB-D sensor.

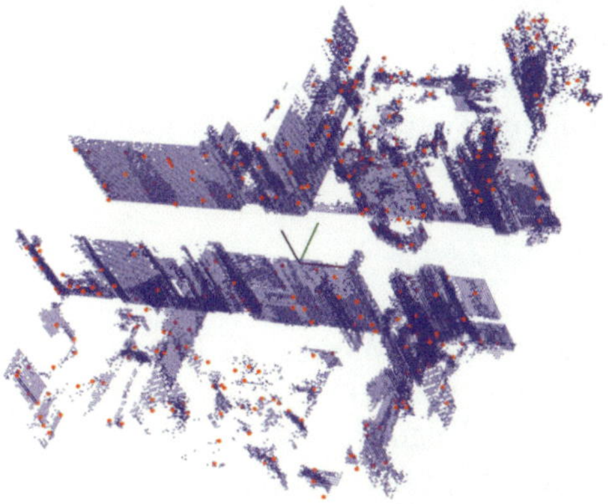

Figure 5.
Points on the example map, for which descriptors were calculated.

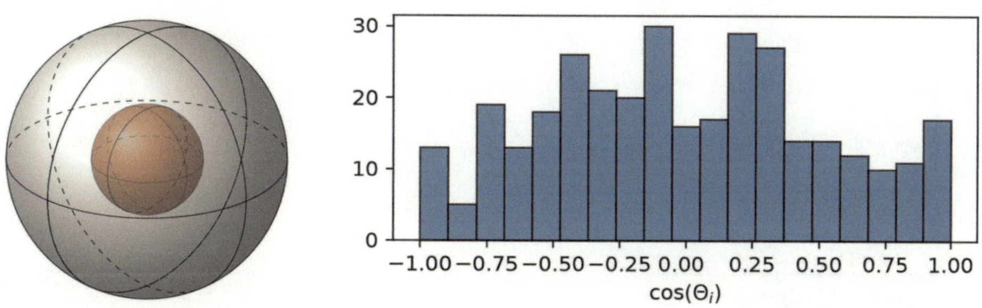

Figure 6.
The spherical support of SHOT descriptor divided into 32 spatial parts and histogram of $\cos\theta_i$.

2.5 Feature matching

To estimate the transformation between maps, it's necessary to find similar features in two maps and matched them to each other. The first feature descriptors set D_M is created for the keypoints of the model $M' = \{m_i | m_i \in \mathbb{R}^3, i = 1, \dots, N_M\}$ and the second set D_s for the scene $S' = \{s_i | s_i \in \mathbb{R}^3, i = 1, \dots, N_S\}$. During the matching process, for each keypoint from the set M', it should be found the point in S' with a similar feature descriptor. Finally, it is calculated a transformation that minimizes distances between pairs of descriptors.

Two matching methods were used. The first is a randomized algorithm SAC (Sample Consensus) [30]. The idea behind it is to use a random search of corresponding features in two sets. The algorithm in each iteration consists of three steps:

- Randomly select k points from the set M' and add them to set $P = \{p_i | p_i \in \mathbb{R}^3, i = 1, \dots, k\}$,
- For each $p_i \in P$, find points with similar descriptors in S' and randomly select from them the one that will make a pair with the point from P,
- For each pair of corresponding points, compute the transformation between points and check the error metric.

The second global alignment method is a GCC (Geometry Consistency Clustering) [31]. It uses the correspondences grouping based on the geometrical distances between them (**Figure 7**).

2.6 Local alignment

Finally, to improve the previous solution, the local alignment is applied. For this purpose, the scene is cropped to the size of the model inflated by a distance d_m (**Figure 8**). Then, the ICP-based algorithm is used to correct a transformation between the scene $S = \{s_i | s_i \in \mathbb{R}^3, i = 1, \dots, N_S\}$ and the model $M = \{m_i | m_i \in \mathbb{R}^3, i = 1, \dots, N_M\}$. The method matches one map (scene) S to the second one M called model in a way that minimizes distances between pairs of points

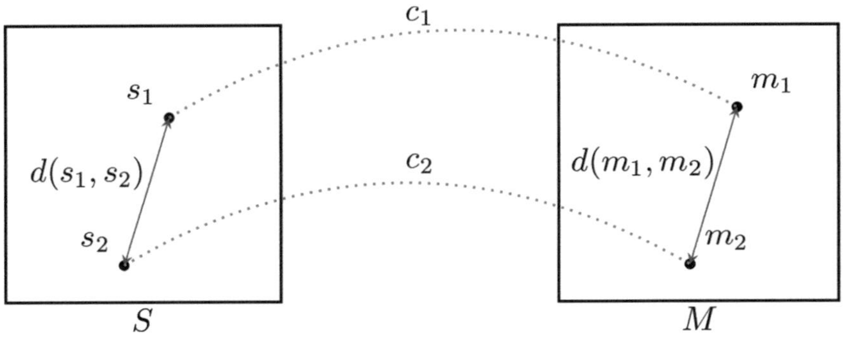

Figure 7.
Correspondences $c_1, \dots, c_n$ grouping in GCC method based on distances between point features in two sets.

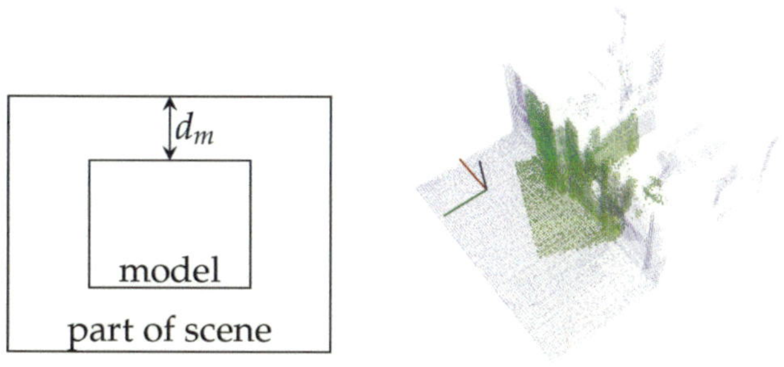

Figure 8.
The scene cropped to the size of the inflated model. On the right side, there is a model matched to the scene.

(nearest neighbors) from both sets. It is done in multiple iterations and a single k-th iteration consists of the following steps:

- $\forall m_i^k \in M$ find the closest point (nearest neighbor) $s_i^k \in S$,
- Minimize distances between corresponding points pairs with the least squares method

$$E(R, t) = \frac{1}{N_M} \sum_{i=1}^{N_M} \left\| Rm_i + t - s_i^k \right\|^2, \tag{4}$$

- Transform model according to estimated rotation R and translation t

$$M^{k+1} = RM^k + t^k, \tag{5}$$

- Terminate if the error value is below the threshold τ.

Also, other methods were used for the local corrections. The first is the OICP (*Occupancy Iterative Closest Point*) which is a variant of ICP that utilizes additional information - occupancy in minimized goal function. Such information is included in occupancy-based maps like octomaps. The second method was the NDT [23] which divides space into voxels and estimates normal distribution parameters in each voxel. Parameters of the normal distributions are calculated with Eqs. (1 and 2).

2.7 Transformation evaluation

The solutions have been evaluated with a computed fitness score

$$f_s = \begin{cases} \sum_{i=1}^{n} \frac{\left\| p_i - q_i \right\|_2}{n_w} w_i, & \text{if } n_w \geq n_{th} \\ +\infty, & \text{otherwise} \end{cases}. \tag{6}$$

It represents the mean error between n pairs of corresponding points (p_i, q_i) from two data sets $P = \{p_i | p_i \in \mathbb{R}^3, i = 1, \ldots, n\}$ and $Q = \{q_i | q_i \in \mathbb{R}^3, i = 1, \ldots, n\}$. Nevertheless, the basic fitness score that is a mean distance is not very robust, for instance, it can be calculated on the basis of a small number of pairs. Because of that, binary weights

$$w_i = \begin{cases} 1, & \text{if} \left\| p_i - q_i \right\| \leq d_{th} \\ 0, & \text{otherwise} \end{cases}, \tag{7}$$

were calculated for each pair depending on the maximum distance between corresponding points d_{th}. Due to that, only pairs with smaller distances are considered in calculations. Also, if the number of pairs with a positive weight

$$n_w = \sum_{i=1}^{n} w_i \tag{8}$$

is below the threshold value n_{th}, the fitness score is set to infinity. The presented metric avoids false good nodes matching because of the use of a threshold. Without it, it is possible to get a low fitness score only based on a small number of points pairs.

3. Validation

The presented maps integration method has been validated in numerous experiments. The first set of experiments was based on data from public datasets from Freiburg University. Another experiment contains data from two Tuerlebot robot runs. Finally, different map alignment methods were compared.

3.1 Experiments with public datasets

The dataset prepared at Freiburg University has been released under the *Creative Commons Attribution License CC 3.0* [32]. It contains maps created with a 2D SICK LMS laser scanner that was placed on a pan-tilt unit to get the additional dimension. As a result, the accuracy of maps is quite better than the accuracy of maps created with the RGB-D sensor but they do not provide information about the surface color.

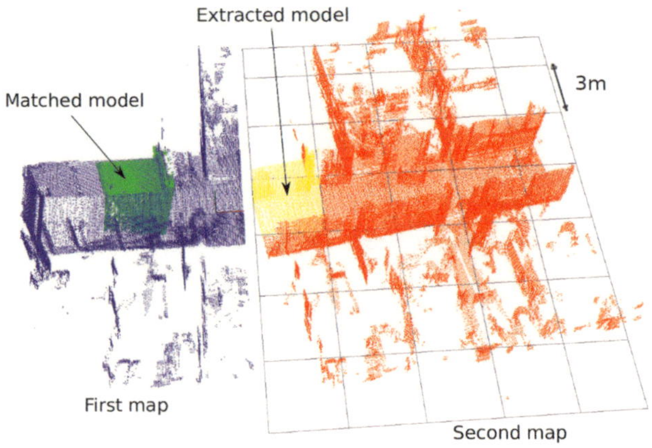

Figure 9.
Global alignment example. A model (yellow) has been extracted from the one map (red) and it was matched to the second map (blue).

Figure 9 shows the example of the global alignment process. The selected results of the maps integration have been placed in **Figures 10** and **11**. Other results are shown in **Table 1**, where:

- n_1 and n_2 are sizes (a number of nodes) of input maps,
- T_R is a real transformation between maps in format $(x, y, z, roll, pitch, yaw)$,
- r is an approximated size of an overlapping area of both maps as a percentage of the full map,
- f_s is a fitness score,

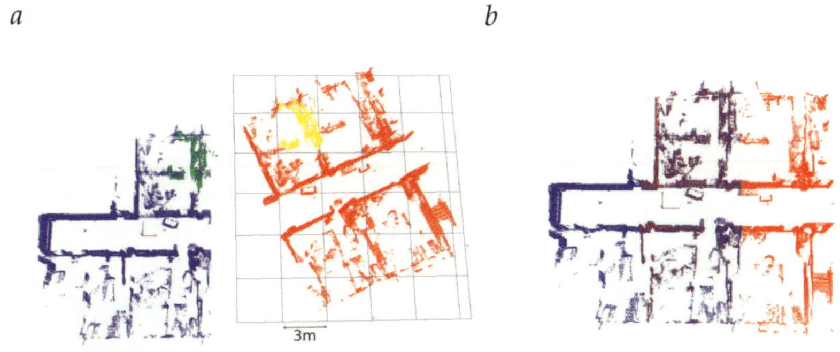

Figure 10.
Indoor maps integration (a) and an integrated map (b).

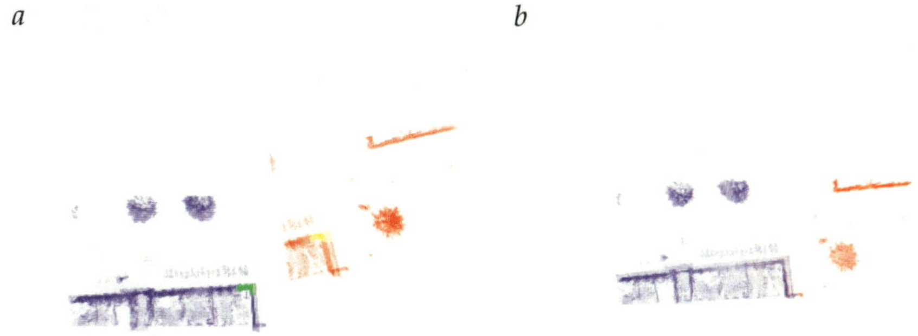

Figure 11.
Outdoor maps integration (a) and the integrated map (b).

n_1	n_2	T_R (x, y, z, R, P, Y)	r	f_s [m]	T_{err}	t [s]
51k	67k	(10, 1.5, 0.1, 3°, 2°, 25°)	12%	0.02	0.6	4.3
		(10, 1.5, 0.1, 3°, 2°, 25°)	24%	0.002	0.16	2.1
		(10, 1.5, 0.1, 0,0,0°)	36%	0.0006	0.10	1.9
		(12, 6, 0.5, 5°, 5°, 60°)	24%	0.0009	0.21	2.4

Table 1.
Results of transformations estimation.

- T_{err} is an error value between real and estimated transformation and is calculated as matrix norm $T_{err} = \|T_{est} \cdot T_R^{-1} - I_4\|_F$,
- t is a processing time in seconds.

3.2 Experiment with turtlebots robots

To validate the integration algorithm with data captured with RGB-D sensor, the experiments with two mobile robots Turlebots (**Figure 12**) were carried out. The Turtlebots were equipped with an odometry system, lidar Hokuyo UST-10LX and RGB-D sensor Intel RealSense D435.

The system used for the octomaps creation (**Figure 13**) was built upon the ROS framework and the GMapping SLAM. During the online session, the 2D SLAM

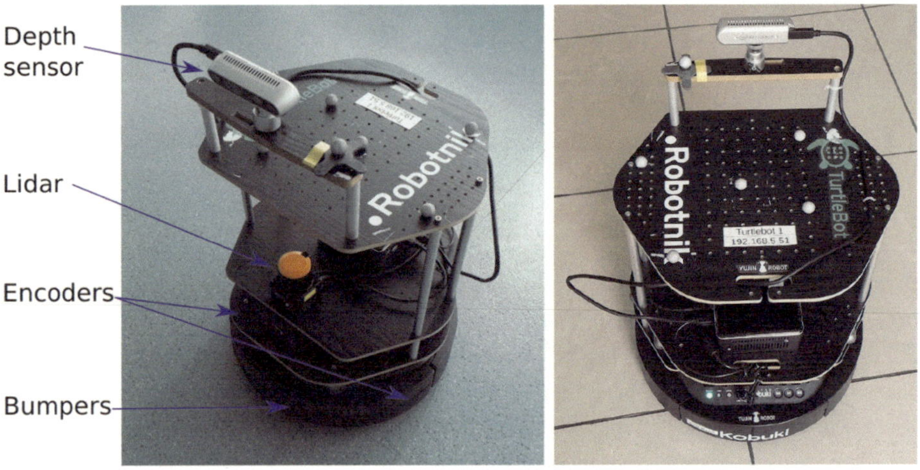

Figure 12.
Turtlebot robot equipped with multiple sensors used for localization and mapping.

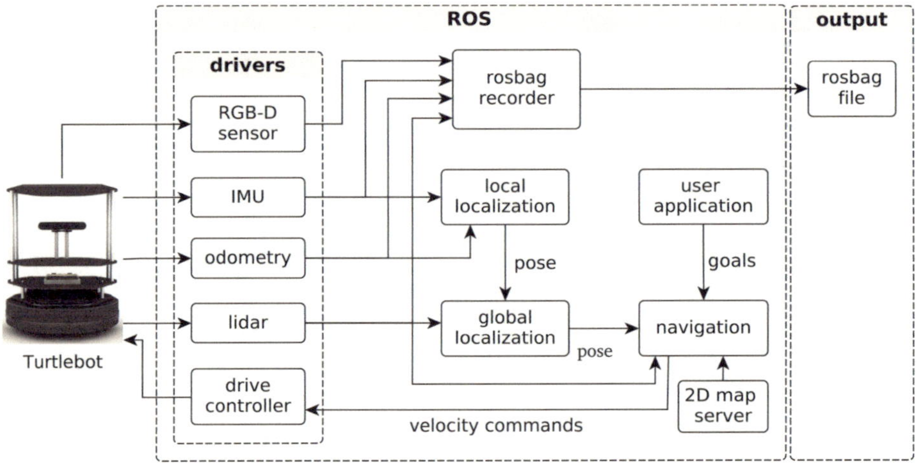

Figure 13.
The high-level control system of the mobile robot used to create the octomap.

algorithm estimated robot pose based on data from IMU, lidar, and odometry. On the other hand, the offline 3D map creation (octomap) used data from RGB-D sensor and pose estimated by SLAM.

The experiments with mobile robots were carried out in a robotics laboratory and corridor localized on the campus of Wrocław University of Science and Technology (**Figure 14**). Experiments results have been shown in **Figures 15** and **16**.

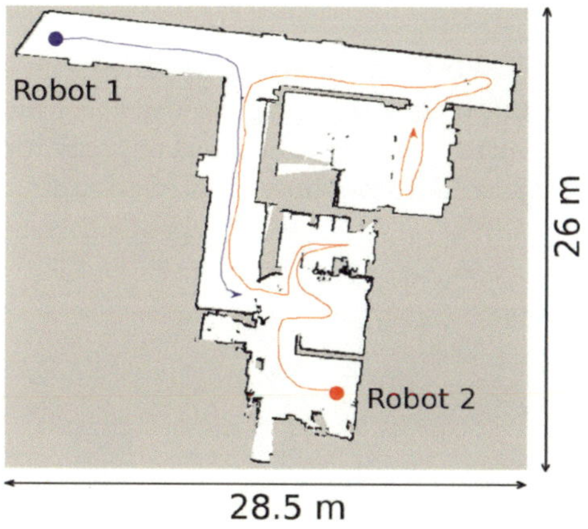

Figure 14.
The paths of two robots followed during the experiment. The paths are presented on the 2D map created for the same localization with lidar and GMapping SLAM.

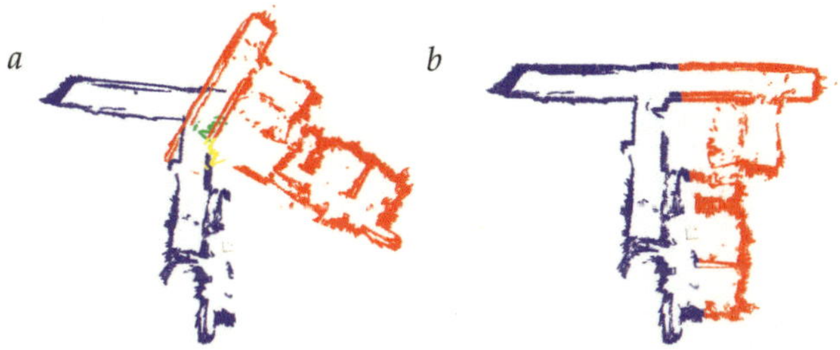

Figure 15.
An experiment with a mapping of the laboratory in WrocÅ‚aw University of Science and Technology. The figure shows maps before merging (a) with extracted (yellow) and matched model (green). On the other side, there is a merged map (b).

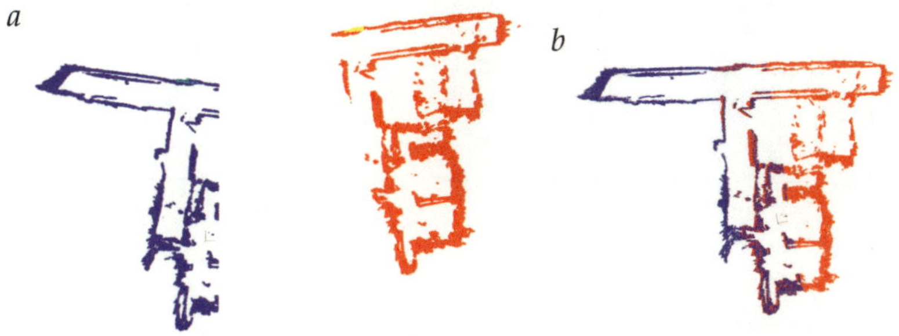

Figure 16.
Integration of maps from the same location but with different initial poses of robots.

3.3 Comparison of the alignment methods

The performance of alignment algorithm variants has been evaluated in multiple testing cases which contain distinctive maps. The mean value T_{err} has been calculated for all testing cases. Additionally, it was verified if a division of one map into models (model extraction process) could speed up the alignment procedure. Therefore, there are the comparison of two kinds of methods, with *DIV* postfix and without it. Added postfix means that method uses a model extraction procedure.

The diagram (**Figure 17**) shows mean errors for different local alignment and global alignment methods checked separately. On the other hand, the diagram (**Figure 18**) contains mean errors for combinations of local and global alignment methods. Additionally, **Figure 19** shows how map overlapping percentage affects the alignment error.

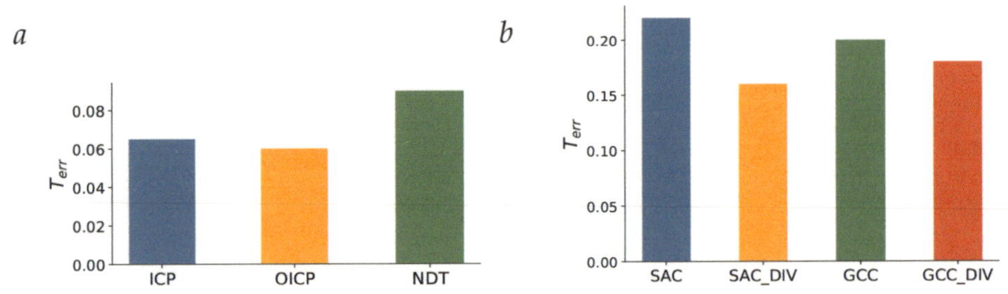

Figure 17.
Mean error between real and estimated transformations for (a) local alignment methods, and global alignment methods.

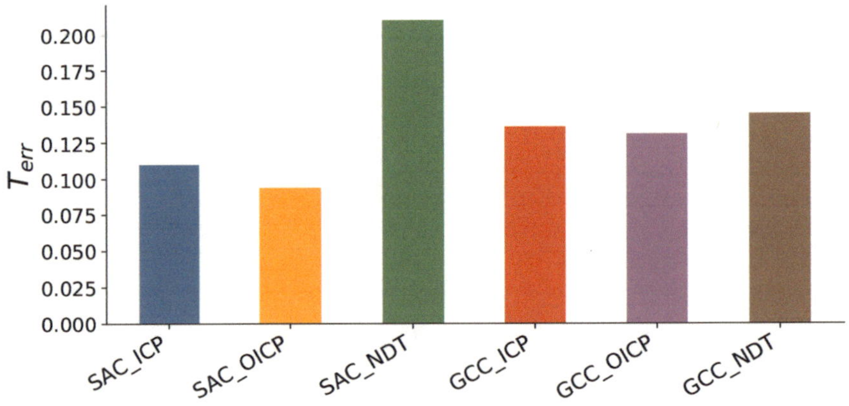

Figure 18.
Mean error between real and estimated transformations for combination of global, and local alignment methods.

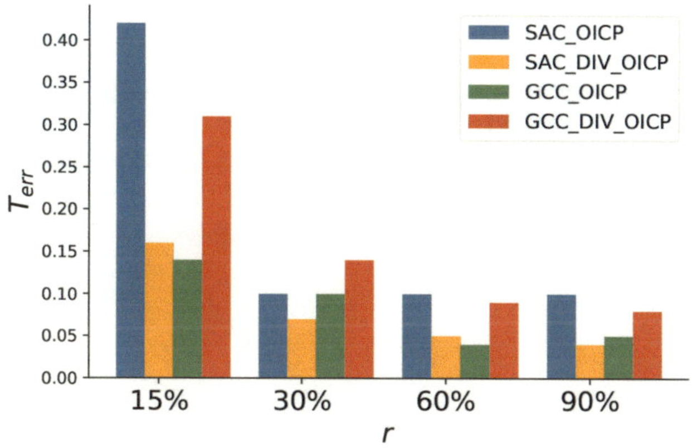

Figure 19.
Comparison of mean error depending on maps overlapping percentage.

4. Conclusions

The chapter presents a 3D maps integration algorithm that does not need initial knowledge about the relative poses of robots. Instead, it uses the feature extraction and matching idea. Therefore, the algorithm needs an overlapping area between maps to integrate them successfully.

The algorithm validation is based on public datasets with octomaps and experiments with two mobile robots. Multiple variants of the method have been validated and compared, especially different combinations of local and global alignment methods, but also additional model extraction procedure.

The results show that the approach is effective in various environments, especially if maps have at least 15% of common area. The best results have been noticed for probabilistic SAC method with local alignment OICP and enabled models extraction procedure.

On the other hand, the maps integration method has some limitations. The computational cost of the data processing is significant, especially the cost of the global alignment step that is necessary always during the first maps exchange between specific pair of robots. The whole integration process takes about 5 seconds in the case of two maps with sizes about $20m \times 20m \times 2m$ and resolution equal to 5 cm. It has been also noticed that it is hard to find a transformation between maps that contain a ground plane because the ground plane does not contain enough distinctive features. Nonetheless, it can be easily improved, by removing the ground plane from maps. The scaling also needs some attention, as the method was already tested only with two robots.

A significant drawback of the approach that is still not addressed is the loop closure problem. In the current version, it was assumed that local alignment errors are small enough and their influence is negligible. However, it is not true and multiple maps integrations lead to increased errors, what has an impact on the quality of the output map. This issue may be solved by providing a high-level graph-based approach with graph optimization to manage multiple partial maps from robots.

Acknowledgements

This research was supported by the National Science Centre, Poland, under the project number 2016/23/B/ST7/01441.

Author details

Michał Drwiega* and Elżbieta Roszkowska
Department of Cybernetics and Robotics, Wrocław University of Science and Technology, Wrocław, Poland

*Address all correspondence to: drwiega.michal@gmail.com

References

[1] Chang Y, et al. LAMP 2.0: A Robust Multi-Robot SLAM System for Operation in Challenging Large-Scale Underground Environments

[2] Ferrein A, Scholl I, Neumann T, Krückel K, Schiffer S. A system for continuous underground site mapping and exploration. In Mahmut Reyhanoglu and Geert De Cubber, Unmanned Robotic Systems and Applications. London, United Kingdom: IntechOpen; 2019. Available from: https://www.intechopen.com/chapters/67435

[3] Yu S, Fu C, Gostar AK, Hu M. A review on map-merging methods for typical map types in multiple-ground-robot SLAM solutions. Sensors. 2020; **20**(23):6988

[4] Andersone I. The characteristics of the map merging methods: A survey. Scientific Journal of Riga Technical University. Computer Sciences. 2010; **41**(1):113-121

[5] Lee H-C, Lee S-H, Lee T-S, Kim D-J, Lee B-H. A survey of map merging techniques for cooperative-SLAM. In: 2012 9th International Conference on Ubiquitous Robots and Ambient Intelligence (URAI). Vol. 1. Daejeon, Korea (South): IEEE; 2012. pp. 285-287

[6] Magnusson M, Vaskevicius N, Stoyanov T, Pathak K, Birk A. Beyond points: Evaluating recent 3D scan-matching algorithms. In: 2015 IEEE International Conference on Robotics and Automation (ICRA). Vol. 1. Seattle, WA, USA: IEEE; 2015. pp. 3631-3637

[7] Konolige K, Fox D, Limketkai B, Ko J, Stewart B. Map merging for distributed robot navigation. In: Proceedings 2003 IEEE/RSJ International Conference on Intelligent Robots and Systems (IROS 2003) (Cat. No.03CH37453). Vol. 1. Las Vegas, Nevada, USA: IEEE; 2003. pp. 212-217

[8] Lee HS, Lee KM. Multi-robot SLAM using ceiling vision. In: 2009 IEEE/RSJ International Conference on Intelligent Robots and Systems. Vol. 1. 2009. p. 6

[9] Tungadi F, Lui WLD, Kleeman L, Jarvis R. Robust online map merging system using laser scan matching and omnidirectional vision. In: 2010 IEEE/RSJ International Conference on Intelligent Robots and Systems. Vol. 1. Taipei: IEEE; 2010. pp. 7-14

[10] Wang K, Jia S, Li Y, Li X, Guo B. Research on map merging for multi-robotic system based on RTM. In: 2012 IEEE International Conference on Information and Automation. Vol. 1. Shenyang, China: IEEE; 2012. pp. 156-161

[11] Besl P, McKay N. A method for registration of 3-D shapes. IEEE Transactions on Pattern Analysis and Machine Intelligence. 1992;**14**(2): 239-256

[12] Han J, Yin P, He Y, Gu F. Enhanced ICP for the registration of large-scale 3D environment models: An experimental study. Sensors. 2016;**16**(2):228

[13] Carpin S. Fast and accurate map merging for multi-robot systems. Autonomous Robots. 2008;**25**(3): 305-316

[14] Saeed Gholami S, Magnusson M. 2D map alignment with region decomposition. Autonomous Robots. 2019;**43**(5):1117-1136

[15] Saeedi S, Trentini M, Seto M, Li H. Multiple-robot simultaneous localization and mapping: A review: Multiple-robot

simultaneous localization and mapping. Journal of Field Robotics. 2016;**33**(1):3-46

[16] Surmann H, Berninger N, Worst R. 3D mapping for multi hybrid robot cooperation. 2017 IEEE/RSJ International Conference on Intelligent Robots and Systems (IROS). 2019;**1**: 626-633

[17] Yue Y, Yang C, Wang Y, Senarathne PGCN, Zhang J, Wen M, et al. A Multilevel fusion system for multirobot 3-D mapping using heterogeneous sensors. IEEE Systems Journal. 2019;**1**:1-12

[18] Drwięga M. Features matching based merging of 3D maps in multi-robot systems. In: 2019 24th International Conference on Methods and Models in Automation and Robotics (MMAR). IEEE. pp. 663-668

[19] Jessup J, Givigi SN, Beaulieu A. Robust and efficient multi-robot 3D mapping with octree based occupancy grids. In: 2014 IEEE International Conference on Systems, Man, and Cybernetics (SMC). San Diego, CA, USA: IEEE; 2014. pp. 3996-4001

[20] Hornung A, Wurm KM, Bennewitz M, Stachniss C, Burgard W. OctoMap: An efficient probabilistic 3D mapping framework based on octrees. Autonomous Robots. 2013;**34**(3):189-206

[21] Wurm KM, Hornung A, Bennewitz M, Stachniss C, Burgard W. OctoMap: A Probabilistic, Flexible, and Compact 3D Map Representation for Robotic Systems. ICRA; 2010. p. 8

[22] Bonanni TM, Corte BD, Grisetti G, et al. IEEE robotics and automation letters. 2017;**2**(2):1031-1038

[23] Magnusson M, Lilienthal A, Duckett T. Scan registration for autonomous mining vehicles using 3D-NDT. Journal of Field Robotics. 2007; **24**(10):803-827

[24] Magnusson M, Nuchter A, Lorken C, Lilienthal AJ, Hertzberg J. Evaluation of 3d registration reliability and speed - A comparison of ICP and NDT. In: 2009 IEEE International Conference on Robotics and Automation. Kobe, Japan: IEEE; 2009. pp. 3907-3912. Available from: https://ieeexplore.ieee.org/document/5152538

[25] Censi A, Carpin S. HSM3D: Feature-less global 6DOF scan-matching in the Hough/Radon domain. In: 2009 IEEE International Conference on Robotics and Automation. Kobe: IEEE; 2009. pp. 3899-3906

[26] Chen S, Nan L, Xia R, Zhao K, Wonka P. Plade: A plane-based descriptor for point cloud registration with small overlap. IEEE Transactions on Geoscience and Remote Sensing. 2020; **58**(4):2530-2540

[27] Drwięga M. 3D Map Server Software. Available from: https://github.com/mdrwiega/3d_map_server

[28] Zhong Y. Intrinsic shape signatures: A shape descriptor for 3D object recognition. In: 2009 IEEE 12th International Conference on Computer Vision Workshops, ICCV Workshops. Kyoto, Japan: IEEE; 2009. pp. 689-696

[29] Tombari F, Salti S, Di Stefano L. A combined texture-shape descriptor for enhanced 3D feature matching. In: 2011 18th IEEE International Conference on Image Processing. Brussels, Belgium: IEEE; 2011. pp. 809-812

[30] Rusu RB, Blodow N, Beetz M. Fast point feature histograms (FPFH) for 3D registration. In: 2009 IEEE International

Conference on Robotics and Automation. Kobe: IEEE; 2009. pp. 3212-3217

[31] Chen H, Bhanu B. 3D Free-Form Object Recognition in Range Images Using Local Surface Patches. p. 4

[32] A. Hornung. OctoMap 3D Scan Dataset. Available from: http://ais.informatik.uni-freiburg.de/projects/datasets/octomap/

Chapter 6

Scalable Algorithms for Simultaneous Mapping and Localization of Mobile Robot Swarms

Anton Filatov and Kirill Krinkin

Abstract

The chapter is devoted to the development of scalable algorithms for multi-agent solution of the SLAM problem. These algorithms are applicable to robots with limited computational resources, having limited computational power and memory, small spatial size, and power from a portable battery. To simplify the description, only robots equipped with LIDAR are considered. The main focus is as follows: a scalable multi-agent SLAM algorithm based on Dempster-Shafer theory; an algorithm for filtering two-dimensional laser scans to free up computational resources; evaluation of the accuracy of the map and trajectory constructed by the multi-agent algorithm; performance evaluation on resource-limited computing devices.

Keywords: autonomous systems, mobile robots, artificial intelligence, localization, SLAM

1. Introduction

Service robots become more and more common every year. There are a lot of areas where they can be applied: medicine, i.e., COVID robots, delivery, rovers, etc. One of the tasks that a service robot faces is an orientation on the environment. The accuracy of the map in the onboard computer's memory, according to which the robot is moving, determines how precisely it will move. The determination of the robot's own position on this map is also significant. Usually it is necessary to determine its own position to an accuracy of centimeters, no satellite location system is capable of it.

To ensure accurate localization, algorithms that solve the problem of SLAM (Simultaneous Localization And Mapping) are used. If it is known that the environment may dynamically change, then the most reasonable approach is to build a map online instead of using a pre-constructed one. For example, a car driving on public roads must avoid potholes on the road or avoid temporarily blocked sections of road. Equipping an unmanned car with such information in advance is a challenge. So using a pre-constructed map is certainly helpful, but it is also necessary to update the map as you drive.

Using multiple agents working together and building a map of the environment together allows for both faster map building and more accurate localization for each robot individually. The acceleration is achieved due to the fact that the researched area is divided into sections, and each section is studied by only a subgroup of agents, and then all the studied sections of the map are combined. The increase in accuracy compared with the single-agent algorithm is achieved due to the fact that each agent not only complements but also verifies the map of other agents. This allows correcting possible errors that occur in calculating its own position and constructing the map on the fly.

The **goal** of this work is to develop scalable algorithms for multi-agent SLAM, applicable to robots with limited computational resources. To increase the accuracy of the map construction, it is proposed to use **the Dempster-Shafer theory** [1] instead of the classical Bayesian one. Experiments have shown that the use of this theory increases the accuracy of the SLAM algorithm. In addition, the developed algorithms can be applied to robots with limited computational resources. In the context of this paper, a robot is a mobile platform equipped with a computing device and a LIDAR—a sensor that can measure the distance to obstacles surrounding the robot. Resource constraints are applied to the computing device, the applicability of different LIDARs is not analyzed in this paper. Thus, a robot with limited computational resources is a robot with limited computational power and RAM, small size, and power from a portable battery. At the moment, typical representatives of such devices are Raspberry Pi or Jetson Nano products.

2. Swarm

First of all, it is necessary to introduce the notion of "swarm," which is the executor of the multi-agent SLAM algorithm. A swarm of robots is a set of autonomous robots with comparable technical equipment for observing the environment in a limited area, they have no hierarchy and exchange information with each other to mark the environment together. Therefore, each robot in the swarm performs the same functions, there is no vertical hierarchy in the swarm. There is no central node—the server, whose absence could cause the entire system to stop working.

There are various papers describing the solution to the SLAM problem, both for a single robot and for a swarm. For example, in the works [2–4], the single-agent algorithms that are popular at the time of these works are listed and compared. There are algorithms that were directly developed as multi-agent algorithms. Usually their peculiarity is that there is some hierarchy and distribution of roles in the swarm. However, there are approaches where robots are independent agents and exchange data with each other asynchronously. In addition, there are a number of studies where single-agent pipelines were augmented with communication modules, and thus, the single-agent algorithm was "extended" to the multi-agent case. More details about the overview of such approaches are written in the papers [5–7]. In the current paper, it is considered the latter approach of those described above.

Each agent in the swarm independently determines the moment when it is necessary to start exchanging information with other agents. Naturally, the bases for such a decision are the same for all members in the pack. The model of this behavior for two robots in the swarm is presented in **Figure 1**.

The central node here is the communication module, which can perform a combination of its own data with another agent data. If communication with other robots in the swarm is not required, the communication module processes only its own data.

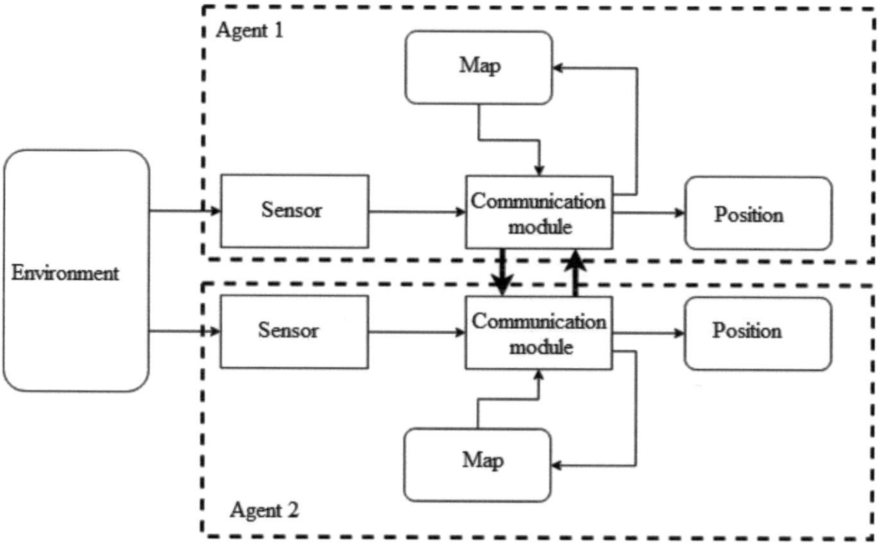

Figure 1.
The model that describes the considered multi-agent SLAM algorithm.

3. Algorithm description

The developed algorithm will be used by the agents in an indoor environment. This means that the size of the studied space does not exceed several hundred square meters. In this case, it is impossible to use GPS, because the error of positioning using GPS is several meters, which is unacceptable. One can use a LIDAR as a sensor, the effective radius of which is measured in tens of meters. Choosing a LIDAR as a sensor allows you to build a map in the form of a grid of occupancy. Thus, the map constructed as a result of the algorithm will be a plan of the building on which the agents move.

Important fact is that the problem of controlling the mobile platform to explore the environment is not considered. The task of the developed algorithm includes only the processing of measurements obtained from the LIDAR and the odometry sensor and the construction of the map.

The indoor environment in which agents move should not contain dynamically changing areas. The appearance of rare and small noises is possible (e.g., the movement of people or other mobile platforms). However, moving furniture or columns in the room, especially at times when the agent is not observing the relevant part of the environment, is not allowed.

It is mandatory that all agents executing the SLAM algorithm must be equally equipped: not only have the same computing capabilities, but also the same mobile platform and the same LIDARs. This requirement is due to the fact that each agent executes the same algorithm, so the output data structure of the agents must be the same. If, for example, the computing power of one of the agents is greater than that of the other, the quality of the map generated by the agent with less power will be worse due to the lack of time for data processing. In this case, the map obtained as a result of the merging will not only contain areas with different degrees of clarity, but if some areas intersect, the less accurate map can introduce a significant error in the result.

3.1 Description of the features of the multi-agent algorithm core

In order for the algorithm to work robustly, it is necessary to get laser scan data and odometry data as inputs. It has already been said that odometry is not a mandatory parameter; however, it serves to increase accuracy.

Odometry (as a prior estimation of position), together with the new laser scan and the currently constructed map, is passed to the input of the scan matcher [8].

The task of this module is to calculate the difference between a prior position estimation and the true position. The true position is the one from which a given laser scan can be captured. A laser scan may contradict the map due to errors of the scan matcher in previous steps or due to an incomplete map; the goal is to compute the most likely scan position.

To calculate the most likely position, one can use a Bayesian approach to probability calculation. The probability of a position is then calculated as the average sum of the probabilities of all scan points overlaid on the map from a given position. Each point of the scan denotes an obstacle, and under ideal conditions each point should be located in an occupied cell of the map. However, it is not enough to introduce a binary division into occupied and unoccupied cells. In reality, a map cell may be too large, and the scan points may be placed in it, as shown in **Figure 2**.

It is logical to assume that the cell has some probability of being occupied. This probability is based not only on the fact that at least one point of the laser scanner hits the cell, but also on how many beams from the LIDAR can pass through the cell before hitting an obstacle. This cell structure opens up the possibility of calculating the probability of position. The position probability can be calculated as the average sum of the probabilities of the cells in which the laser scanning points hit:

$$p_{pose}(scan) = \sum_{p \in points} o_p \cdot \omega_p \tag{1}$$

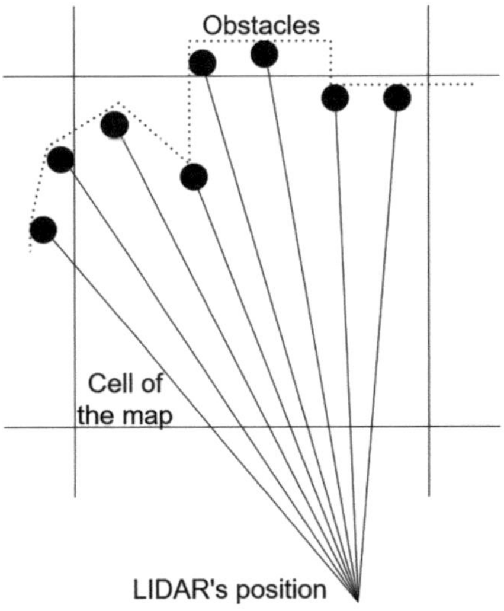

Figure 2.
An example of a part of a laser scan. A lot of scanning points fall on a cell, but such a cell cannot be considered occupied.

where p is a scan point, o_p is an occupancy of the point p, ω_p is a weight of the scan point.

The weight of the scan point is introduced to make the algorithm more stable in corridor conditions. In this case, different robot positions along the corridor direction have very close probability values. Differences in the values of probabilities are introduced by scan points that are different from the corridor walls. Usually such points are located in the direction of movement of the mobile platform. Consequently, laser scanning points located directly opposite the laser rangefinder have a greater weight.

The scan matcher's goal is to calculate the position with the highest probability (when the laser scan points released from that position hit the squares on the map that have the highest probability of being occupied).

The approach to calculating such a position can be arbitrary. In this paper, the Monte Carlo scan matcher [9] is taken as the basis, the idea of which is a stochastic search of positions. Once a position with higher probability than all previous positions is found, the search begins in a lower radius around that position. This process is shown in **Figure 3**.

Figure 3 shows the sequence of steps. Position 1 is a prior estimate of the position. With some radius around it, the positions are being chosen stochastically and their probabilities are calculated. The enumeration is carried out until position 2 with a higher probability than the probability of position 1 is found. The radius around position 2 is smaller than the radius around position 1. Then the operation is repeated several times. Practical application of this algorithm shows that it is sufficient to set a threshold for the number of calculations of position probabilities in general, instead of for the number of such positions. A guarantee of finding a solution by such an algorithm exists only if there are no distinct local maxima of position probabilities in the stochastic search area. Otherwise, it is necessary to increase the search radius, as well as the number of positions to search.

3.2 Applying Dempster-Shafer theory to increase the accuracy of the algorithm

According to the Bayesian approach, each cell in the map contains a number from 0 to 1 that determines the probability of a cell of being occupied. To calculate this

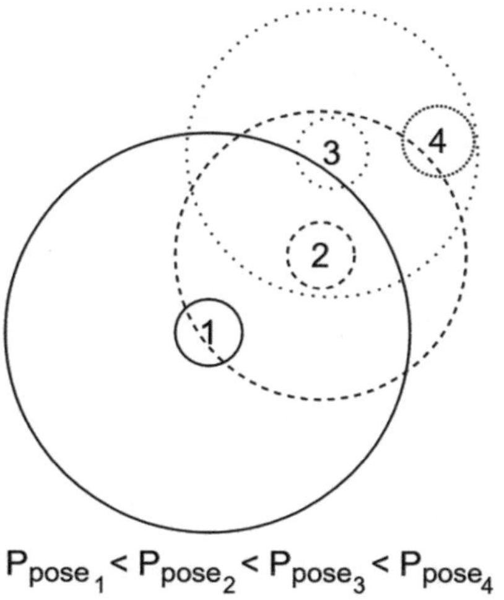

Figure 3.
The visualization of an iterative algorithm for stochastic search for a position with the highest probability.

probability, it is needed to estimate the position of specific scan points in the cell relative to the point from which the laser scan was observed. Each cell on the map has probability p of being occupied and probability q of being free. These probabilities are related by the equation p + q = 1. However, under real-world conditions, the "unknown" state should also be introduced. The cell may have such a state when it is in an unexplored region. Also, the cell may be in the "unknown" state when two consecutive scans contradict each other.

Thus, a cell is described by three probabilities: the probability of being occupied p, the probability of being free q, and the probability of being unknown u. These three probabilities are related by the identity p + q + u = 1. Unlike the previous approach, where one probability is expressed through another, this approach allows each probability to be expressed through a pair of others. The question arises how to combine cells following these rules.

To answer this question, the Dempster-Shafer theory is applied [1, 10]. To strictly follow this theory, it is also necessary to introduce a state of conflict in the cell of the map. From a natural point of view, conflict is the state of a cell not being in any of the possible states. But in reality, no cell in the map can be in a state of conflict, which means that this probability must be distributed among the other states. It should be noted: if a cell is in an unknown state, then it is simultaneously in free and occupied states.

In the context of the problem at hand, the Dempster-Shafer theory is applied during the scan matcher, when it is necessary to find out the probability of a map cell being occupied. Immediately after the scan matcher runs, the second step of applying this theory is to place the probabilities of being free and occupied into the empty map cells that have appeared in the LIDAR observation area. To do this, the rule of combining probabilities, described by the equation below, is applied.

$$m_{1,2} = \frac{1}{1-K} \sum_{B \cap C = A \neq 0} m_1(B) \cdot m_2(C) \tag{2}$$

where m is the probability of the state, A,B,C are the different states: whether the cell is free, occupied, or unknown (free or occupied), K is calculated by the formula

$$K = \sum_{B \cap C = 0} m_1(B) \cdot m_2(C) \tag{3}$$

3.3 Description of the multi-agent SLAM algorithm

The distinctive feature of the proposed algorithm is that each agent has no knowledge of the other agents at any time. This allows the system to remain operational when one or more agents fail, or when only one agent remains in the system. Information is exchanged between agents only when two agents are in close proximity to each other.

Based on the described methods and hypotheses, one can construct an algorithm that solves the problem for a swarm of robots. The algorithm is an extension of the single-agent laser single-hypothesis algorithm, which uses an occupancy grid as a map. Each cell, in addition to the probability of being free, also contains a probability of being occupied and of being unknown. These probabilities follow Dempster-Shafer theory. The scheme of such a multi-agent algorithm is shown in **Figure 4**.

Consider the algorithm from the point of view of one particular agent, which solves the SLAM problem. The agent possesses only its own computational resources,

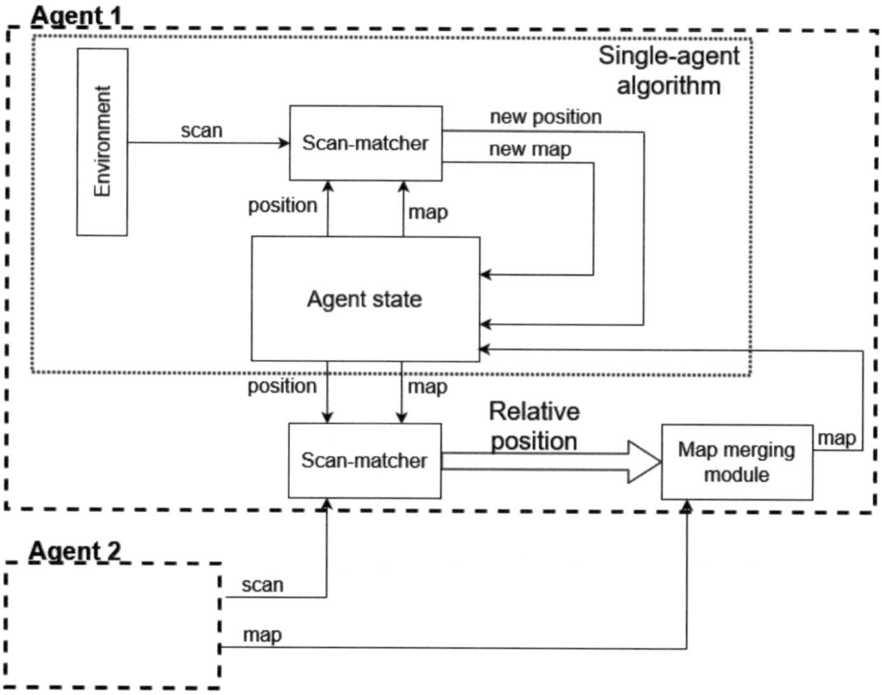

Figure 4.
The scheme of the multi-agent SLAM algorithm.

which are aimed at constructing a map and determining its own position on it. When another agent enters the field of view of the first agent, they begin the process of exchanging all the knowledge accumulated during the past time. It needs to be determined that the other agent has entered the field of view without using sensor data. Otherwise, it would be necessary to label the robot to distinguish it from its environment. And since the algorithm must operate in a situation with no prior knowledge of the environment, the approach that requires labeling the robot is not applicable.

Consequently, it is necessary to determine that one agent is in the direct line of sight of the other by means of some separate sensor. The choice of such a sensor is outside the scope of this work, but as an example, one can use the Bluetooth data transmission technology, the range of which is just a few meters. You can also use the wifi signal and based on the strength of the signal determine the distance between agents. It is possible to equip agents with radio sensors and calculate the distance between them using the radio signal.

When the agents are in close proximity, they can begin the process of exchanging the maps they have built. These maps definitely have a common part, since the data transfer takes place when the agents are almost in the same position and observe the same part of the environment. In addition to the maps, the agents pass each other the current laser scan, which allows applying a scan matcher algorithm that is similar to the one running in each agent's core in a SLAM algorithm.

Thus, an agent receives a laser scan from another agent. Using the scan matcher, it overlays this scan on its own map and determines how far apart the agents are in relation to each other. Consider in more detail the applicability of the scan matcher algorithm to determine such relative position. In general, the scan matcher takes as input the prior agent position, the current laser scan, and the constructed map. The output parameters for calculating relative position are the vector by which a prior position estimate must be changed in order to calculate its posterior estimate. In other words, the scan matcher calculates the difference between the position obtained from the input and the position from which the resulting laser scan can actually be seen, given the resulting map.

This description of the scan matcher assumes that when it receives a laser scan from another agent as input, it will get the distance between the agents as output. The condition that the part of the environment captured on the scan is present in the map must be satisfied. The fulfillment of this condition is guaranteed. The scan matcher algorithm will be discussed in more detail in a further analysis.

After determining the mutual position of the agents, it is possible to start merging the maps. To do this, the second agent passes the first constructed map. Provided both maps obtained are error-free, map merging is a straightforward operation, where it is sufficient to go through each cell of both maps and choose either the largest, smallest, or average occupancy of the corresponding cells. The nature of the algorithm embedded in the core of each agent allows errors to occur during the map construction process that will never be corrected in the future. In particular, a single-agent SLAM algorithm may be not multi-hypothesis nor graph-based. Then the map constructed by one agent or the other may contain errors. Nor is it known whether the same error is contained in only one map or in both maps. To avoid this, the Dempster-Shafer theory is used.

The pseudocode of the multi-agent SLAM algorithm is presented below.

```
function MATCH(observation, pose, map)
    for random position ∈ neighborhood(pose) do
        pose_score ← try_insert(position + observation,map)
    end for
    return position[best_score],best_map
end function
while True do
    pose_i, map_i ← MATCH(observation_i,pose_{i−1},map_{i−1}) ▶ a single SLAM routine
    if another_robot_is_near() then
        foreign_map_i,foreign_observation_i ← receive_foreign_state()
        foreign_pose ← MATCH(foreign_observation_i, pose_i, map_i)
        relative_pose ← pose_i − foreign_pose
        map_i ← combine_with_thDS(map_i,foreign_map_i, relative_pose)
    end if
end while
```

4. Laser scan filtering algorithm

LIDARs, as input data providers, have a common flaw: they collect too little and too much data at the same time. On the one hand, there is little data, because it is impossible to smooth or approximate it without significant loss of accuracy; on the other hand, there is a lot of data, because it takes a lot of memory to store and process data from modern laser scans, which are taken more than 30 times per second [11].

The need to develop a filter for laser scans arises. Normally one does not need to shoot laser scans as often, unless the scanner is mounted on a vehicle traveling at 60 km/h. In this case, the environment can change dramatically in 0.03 seconds. On the other hand, if the robot is moving indoors and has an average velocity of about 0.5–1 m/s [12, 13], such a number of dense point clouds from the laser rangefinder is redundant. The idea of the filter is based on storing several consecutive scans in a sliding window and comparing each new scan with the scans from that window. If each new scan correlates strongly with each scan from the window—it should be discarded.

The basic idea of the developed algorithm is to compare the current laser scan with the previous one. If a new scan is similar to a previous one, it should not be processed; and in order to avoid noise in the observations, the current scan should be compared with several previous scans instead of just one. Thus, a sliding window of scans appears, which is capable of evaluating each new incoming scan.

Usually a laser scan consists of several thousand points. Calculating the correlation of scans by brute force method requires $O(n^2)$ operations, which may exceed a million iterations. To reduce this amount, special points on the scan can be singled out, which would take a lot of time. Instead of using the raw laser scan data to calculate the correlation, it is proposed to build a histogram for each scan.

Consider the method of constructing a division histogram by distance ranges. For each scan, the maximum and minimum values of the distance to obstacles are known. Consequently, it is possible to divide this range spread into several interval, and then calculate the number of points corresponding to each interval. Two consecutive scans usually should not differ significantly, so the distance histograms should be close to each other. In practice, if the robot is not rotating, the difference between the two scans is insignificant, and the values of each column of the histogram change little. If the robot is rotating, the difference is more significant. However, you can update the approach: instead of calculating the number of points in each column, you can calculate the median value of distances for each interval. In this way, the two consecutive histograms become more diverse.

4.1 Correlation criteria

The next step after creating the histograms of each scan is to calculate their correlation. Here one can use the methods of mathematical statistics and consider the histogram as a random variable with an unknown distribution. Since all the histograms from the window are generally similar to each other, it is assumed that the distribution is the same.

There are several well-known ways to calculate the correlation of random variables: Pearson correlation [14], Spearman correlation [15], and Kendall correlation [16]. The simplest is the Pearson correlation coefficient. It is calculated by the equation

$$P_{X,Y} = \frac{cov(X, Y)}{\sigma_X \sigma_Y} \tag{4}$$

where X,Y are some random variables, cov is covariance of random variables, σ is the variance of a random variable.

If a random variable consists of n observations, and x_i is the observed value of this variable at the i-th step, the Pearson coefficient can be calculated by the following equation:

$$P_{X,Y} = \frac{n\left(\sum_{i=1}^{n} x_i y_i\right) - \left(\sum_{i=1}^{n} x_i\right)\left(\sum_{i=1}^{n} y_i\right)}{\sqrt{\left[n\sum_{i=1}^{n} x_i^2 - \left(\sum_{i=1}^{n} x_i\right)^2\right]\left[n\sum_{i=1}^{n} y_i^2 - \left(\sum_{i=1}^{n} y_i\right)^2\right]}} \tag{5}$$

The value of this coefficient ranges from −1 to 1. A value of 1 is a complete positive linear correlation, 0 is no linear correlation, and minus 1 is a complete negative linear correlation. This correlation is called linear because of the following geometric interpretation. Place the value of the first variable on the abscissa and the value of the second variable on the ordinate of the graph. If the points with the resulting coordinates belong to the same line with a positive derivative, then their correlation is 1. If the derivative is negative, then the Pearson coefficient is minus 1. If no line can be drawn, then the coefficient is 0.

Kendall and Spearman coefficients are used to measure the ordered relationship between two measured variables [17]. It is a measure of rank correlation: the similarity of the rank of the data when ordered by each of the variables. The Spearman coefficient is defined as the Pearson coefficient between ranked variables.

The Kendall correlation coefficient is calculated as follows:

$$P_{X,Y} = \frac{number\ of\ concordant\ pairs - number\ of\ discordant\ pairs}{\binom{n}{2}} \tag{6}$$

To summarize, there are three well-known approaches to calculating correlation. The main drawback of the Kendall coefficient is algorithmic complexity. It requires calculating the rank of a random variable and then calculating the number of matched pairs. In the worst case, this may require $Nlog(N)$ operations. The Spearman coefficient is less complex, but it also requires the introduction of an order relation. Since correlation is computed for histograms of consecutive scans, correctly ranking the values in the histograms is a difficult task. For histograms that are similar in general, every small variation in values must be captured. Consequently, the order function must be sensitive to these fluctuations and at the same time show the true correlation. Therefore, the Pearson correlation coefficient is the most appropriate for the algorithm in question. Its complexity is $O(n)$, it does not require an order function, and it is sufficiently sensitive to fluctuations in the values in the histograms.

4.2 Parameters and constants in the scan filtering algorithm

There are four parameters in the proposed algorithm, the fine-tuning of which must be paid attention to:

- The number of columns in the histogram and, therefore, the number of laser scan points in each column;
- the size of the sliding window;
- P_{pair} intra-window correlation threshold (how much the new scan should correlate with each scan in the window);
- discard threshold – value of total correlation of scan with scans in the window P_{common}.

These four parameters affect the number and nature of the filtered scans. The first is the number of columns in the histogram. All histograms considered have a common feature: the more columns contained in the histogram, the more details are processed for each scan. Consider the LIDAR used in the MIT dataset, which captures laser scans consisting of about 1000 points. Dividing these points into 50 columns means that each group of points contains an average of 20 points. Dividing into 10 columns results in groups of 100 points each.

Despite the intuitive notion that the higher the sensitivity, the higher the accuracy, in real data high sensitivity can be the other way around. For example, if a moving object appears in the field of view of a laser scan, it inevitably leads to a difference in two consecutive scans. Moreover, each sensor contributes noise to the observations, and sometimes (at short distances) this noise can be misinterpreted as a difference between scans. The results of the experiments presented in Section 4.5 show that for 1000 laser scanning points with an angle of view of should be grouped into 15 or 30 columns.

Another important constant is the size of the window in which the previous laser scans are located. The correlation estimate of the current scan is equal to the product of the correlation value of each scan from this window and the current scan. Pearson's correlation coefficient is taken as the correlation value. It is obvious that if the robot with the laser scanner moves fast, then a large number of different scans in the window decreases the final correlation estimate of the new scan. This means that the higher the velocity of the robot, the fewer scans should be stored in the window.

There is the experimentally obtained equation that relates the window size to the average velocity of the robot. The average velocity here is the average distance, in centimeters, that the robot travels between two laser scan images. This equation is a heuristic and allows us to relate the scanning capture property (speed) and the filter property (the amount of information that should be in the window).

$$Window\_size = \frac{27}{v^2} \tag{7}$$

where v is an average velocity in centimeters.

To determine the effect of the window size, it is necessary to consider two parameters closely related to each other and to the window size. The first parameter is the threshold for the Pearson correlation coefficient for each pair of scans. The second is the total correlation coefficient, which is equal to the product of the coefficients. Since the Pearson correlation coefficient is calculated for two consecutive scans obtained with a small time difference, it is obvious that they are highly correlated on average. Therefore, the threshold for a pair of scans should be at least 0.95, or better, 0.98. After calculating the correlation coefficient of a new scan with each scan in the window, all coefficients must be combined. A well-known way to do this is to multiply them. For example, the threshold for a window containing five scans and a pairwise correlation of 0.98 is $0.98^5 = 0.904$.

4.3 Evaluation of the quality and accuracy of the laser scan filter

For testing the scan filter was included in the operation of two SLAM algorithms: vinySLAM [18] and Google Cartographer [19]. The filter determines whether the scan should be processed or discarded before it is passed to the scan matcher of each of the listed algorithms. Consequently, if the scan must be processed, the time required for filtering is added to the total processing time of the scan. Therefore, it is necessary to estimate the algorithmic complexity of the filtering process and then present the

The sequence	vinySLAM	vinySLAM filter	Cartographer	Cartographer filter	% dropped
2011-01-20-07-18-45	0.062 ± 0.004	0.078 ± 0.004	0.131 ± 0.058	0.139 ± 0.041	59
2011-01-21-09-01-36	0.080 ± 0.018	0.089 ± 0.013	0.153 ± 0.072	0.163 ± 0.094	56
2011-01-24-06-18-27	0.096 ± 0.007	0.111 ± 0.013	0.183 ± 0.015	0.181 ± 0.014	59
2011-01-25-06-29-26	0.094 ± 0.006	0.100 ± 0.002	0.176 ± 0.010	0.179 ± 0.012	63
2011-01-27-07-49-54	0.170 ± 0.019	0.121 ± 0.006	0.248 ± 0.014	0.251 ± 0.007	52
2011-03-11-06-48-23	0.534 ± 0.085	0.543 ± 0.034	0.586 ± 0.174	0.642 ± 0.191	58
2011-03-18-06-22-35	0.090 ± 0.020	0.090 ± 0.003	0.130 ± 0.025	0.119 ± 0.017	52
2011-04-06-07-04-17	0.183 ± 0.014	0.213 ± 0.027	0.188 ± 0.011	0.185 ± 0.011	51
2011-01-19-07-49-38	0.305 ± 0.174	0.289 ± 0.181	0.188 ± 0.004	0.189 ± 0.005	50
2011-01-28-06-37-23	0.361 ± 0.175	0.348 ± 0.152	0.378 ± 0.025	0.399 ± 0.030	47

Table 1.
RMSE values and percentage of dropped scans for vinySLAM and cartographer on MIT dataset.

results of applying such a filter to real MIT dataset [20]. It is also required to estimate the proportion of scans that can be discarded without loss of overall accuracy (**Table 1**).

5. Accuracy measures of the multi-agent SLAM algorithm

The data flow exchanged between the main nodes in the developed software can be seen in the sequence diagram shown in **Figure 5**. For simplicity, this diagram shows only the two main nodes, the scan matcher and the detector. The other nodes are omitted, since interaction with them can be considered atomic operations that do not require additional actions. One can also assume that the scan matcher has all the information about the current map, the position of the robot, and the current observation.

In order to assess the accuracy of the SLAM algorithm, it is necessary to quantitatively compare the artifacts obtained during its operation: the map and trajectory of the robot with the true map and trajectory. Quantitative comparison of maps is a task of image analysis. The characteristics of such a comparison are quite comprehensive and require specific knowledge of the true map [21]: scale, permissible error in drawing the image, and others. It is also necessary to calculate what part of the true map the robot had time to observe.

Therefore, to evaluate the quality of the SLAM algorithm, a comparison of the trajectory is used as a sequence of positions obtained during the execution of the SLAM algorithm with the true trajectory. The datasets in question are not accompanied by a true trajectory. However, the MIT University dataset comes with a table that can be used as input to solve the localization problem (in other words, together with a utility that allows you to calculate the trajectory of the robot in the recorded dataset, based on the solution of the localization problem).

Since the true trajectory of the robot for each MIT dataset is known, namely the robot position and the timestamp corresponding to this position, it is possible to

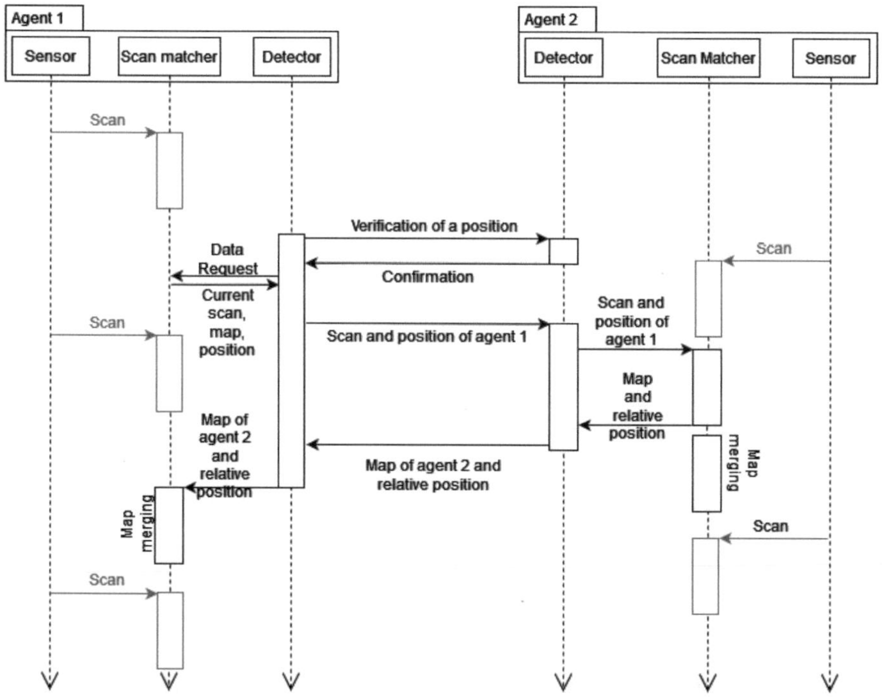

Figure 5.
UML sequence diagram in developed software.

calculate the deviation at each time point. From such an array of deviations, the standard deviation is calculated as a characteristic of the quality of performance of the SLAM algorithm. The disadvantage of the proposed approach is the probability of error in calculating the true trajectory when solving the localization problem. However, it is the most reliable way to quantify the algorithm accuracy.

It is not enough just to calculate the standard deviation from the true trajectory, it is also necessary to determine how small it is. Therefore, a reliable algorithm is needed for solving the SLAM problem and comparing the standard deviation of different algorithms. As an estimating algorithm, gmapping [22] is often used or cartographer. In this paper, the comparison will be made with the cartographer algorithm, as well as with the vinySlam algorithm.

Thus, the first experiment to calculate the accuracy of the multi-agent algorithm consists of the following steps.

1. Run the algorithm on one of the data sequences from the MIT dataset emulating one robot from the swarm.

2. Run the algorithm on another data sequence from the same MIT dataset. You need to make sure beforehand that these sequences are written given the existence of a moment in time when the robots from both sequences are close to each other.

3. At the moment when the robots are close to each other, manually send a command to exchange the current observations and maps.

4. After exchanging data and updating maps, the agents continue executing the algorithm until the end of the recorded data sequence.

5. After completing the data set replay, each agent compares the trajectory it built with the trajectory built by the single-agent algorithm on the same data sequence.

The goal of experiment # 1 is to show that using another robot's map does not reduce accuracy compared with the single-agent algorithm and also allows the agent's map to be supplemented with areas it has not visited so far. This speeds up the work by matching the laser scan to an already constructed map, an algorithmically less costly operation than embedding the scan into an unknown map.

The second experiment consists of running the same sequence twice at the same time. One copy is played without change, and the second copy is played backward. This case guarantees that in the middle of the sequence there will be a point where both agents will be in at the same time. The sequence of steps in experiment 2 is as follows.

1. Run the algorithm on one of the data sequences from MIT.

2. Run the algorithm on the same sequence reproduced in the other direction.

3. When the robots are at the same point, send them a signal to exchange current observations and maps.

4. After exchanging data and updating maps, the agents continue executing the algorithm until the end of the recorded data sequence.

5. When the dataset is complete, each agent compares the trajectory it built with the trajectory built by the single-agent algorithm on the same data sequence.

The key feature of such an experiment is to demonstrate how using exactly the part of the map that the robot will move on in the future affects the accuracy of the results (**Tables 2** and **3**).

The results prove the accuracy of the proposed algorithm, since the RMS error is always less than 0.5 m and is almost always lower than that of the google cartographer graph algorithm. This result is mainly based on the accuracy of the single-agent

The sequence	Trajectory length, m	Multi-agent RMSE, m	Core RMSE, m	Cartographer RMSE, m
2011-01-20-07-18-45	76	0.045 ± 0.005	0.062 ± 0.004	0.131 ± 0.058
2011-01-21-09-01-36	87	0.080 ± 0.018	0.080 ± 0.018	0.153 ± 0.072
2011-01-24-06-18-27	87	0.096 ± 0.011	0.097 ± 0.007	0.183 ± 0.015
2011-01-25-06-29-26	109	0.094 ± 0.009	0.094 ± 0.006	0.176 ± 0.010
2011-01-28-06-37-23	145	0.395 ± 0.190	0.361 ± 0.175	0.201 ± 0.011
2011-01-27-07-49-54	94	0.167 ± 0.018	0.170 ± 0.019	0.248 ± 0.014

Table 2.
RMSE values for experiment # 1 in comparison with RMSE of core SLAM algorithm and Google cartographer.

The sequence	Trajectory length, m	'Forward' RMSE, m	'Backward' RMSE, m
2011-01-20-07-18-45	38 + 24	0.044 ± 0.015	0.021 ± 0.005
2011-01-21-09-01-36	43 + 31	0.080 ± 0.011	0.068 ± 0.012
2011-01-24-06-18-27	43 + 41	0.106 ± 0.009	0.091 ± 0.003
2011-01-25-06-29-26	33 + 61	0.033 ± 0.013	0.097 ± 0.016
2011-01-28-06-37-23 part 1	41 + 39	0.184 ± 0.093	0.231 ± 0.040
2011-01-28-06-37-23 part 2	41 + 39	0.311 ± 0.187	0.272 ± 0.195
2011-01-27-07-49-54	31 + 62	0.142 ± 0.019	0.161 ± 0.022

Table 3.
RMSE values for experiment # 2.

version of the vinySlam algorithm. The largest error can arise if the result of calculating the mutual arrangement of the agents during map merging is a rotation error. Despite the fact that in this case the output map will be consistent, the output trajectory may differ markedly. It is fair to note that this problem occurs in any multi-agent algorithm; and even a graph algorithm can correct for it only under certain conditions, when the agents visit the most inconsistent part of the surrounding world several times.

In addition to estimating accuracy, it is necessary to determine the performance of the multi-agent algorithm on different hardware configurations. The goal of this experiment is to determine the applicability limits on low-performance computing devices. The performance measurements should be divided into two stages:

1. Determination of measurement processing rate during execution of the single-agent part of the algorithm

2. Determination of the rate of data exchange during the encounter, as well as the rate of mutual positioning and map updating

The performance measurements were made on the following configurations.

1. Raspberry Pi 3 Model B (Broadcom BCM2837 processor, 1GB LPDDR2 RAM, Ubuntu Xenial x64 operating system).

2. Raspberry Pi 3 Model B+ (Broadcom BCM2837B0 Quad Core 1.2GHz processor, 1GB LPDDR2 RAM, Ubuntu Xenial x64 operating system).

3. Raspberry Pi 4 Model B (Quad Core 1.5GHz Broadcom BCM2711 processor, LPDDR4 2GB RAM, Ubuntu Xenial x64 operating system).

4. Personal computer (Processor: Intel Core i7-860 4x2.8GHz, DDR3 8GB RAM, Ubuntu Xenial x64 operating system).

The evaluation was performed on Raspberry Pi computers [23], since they are inexpensive and very popular computing devices used in robotics. The experiment was also conducted on virtual machines with resources comparable to those of the Raspberry Pi. The results were too similar, so they are not presented in a separate

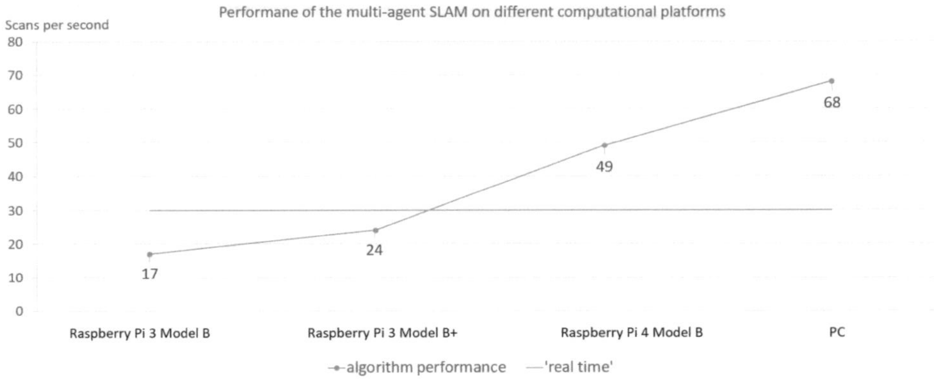

Figure 6.
The amount of processed scans per second in different computational configurations.

graph. Therefore, it can be assumed that other boards with similar resources can be used instead of Raspberry products to obtain performance characteristics.

Figure 6 shows the number of scans per second that can be processed on the listed configurations. As a conditional boundary is taken the input data processing frequency of 30 scans per second, which is comparable with the processing frequency of the human eye. Therefore, one can conventionally consider that if the algorithm is executed with a frequency of more than 30 scans per second, then it works in real-time mode.

6. Conclusion

According to the results of theoretical and experimental studies presented in this paper, the following conclusions can be made.

- Abandoning the graph structure along with the application of Dempster-Shafer theory allows us to achieve good performance of multi-agent SLAM algorithms running on low-powered robots. The rejection of role specialization in a swarm of agents provides good scalability and robustness.

- The application of a filtering algorithm based on the calculation of the correlation of the nearest two-dimensional laser scans allows increasing the frame processing speed up to 40%.

- The conducted experiments show the high accuracy of lightweight algorithms comparable with resource demanding algorithms, such as Google Cartographer. In particular, the root-mean-square error of the multi-agent algorithm was 9.4 cm at a distance of 100 meters covered by one agent, subject to a single synchronization with another agent. The error in determining the position of the robot is comparable to its dimensions.

The application of Dempster-Shafer theory can be promising for data filtering, leader election in graph-based algorithms, and for increasing the accuracy of combining maps with more than two agents.

Acknowledgements

The authors would like to thank Saint Petersburg Electrotechnical University "LETI" and the Pavlov world-class research center in Personalized Medicine, High-Tech Medical Care, and Healthcare Technologies for provided support and materials for working on this paper.

Conflict of interest

The authors declare no conflict of interest.

Author details

Anton Filatov* and Kirill Krinkin
Saint Petersburg Electrotechnical University, Saint Petersburg, Russia

*Address all correspondence to: aifilatov@etu.ru

References

[1] Dempster AP. The Dempster–Shafer calculus for statisticians. International Journal of Approximate Reasoning. 2008;**48**(2):365-377

[2] Huletski A, Kartashov D, Krinkin K. Evaluation of the modern visual slam methods. In: 2015 Artificial Intelligence and Natural Language and Information Extraction, Social Media and Web Search FRUCT Conference (AINL-ISMW FRUCT). Helsinki, Finland. 2015. pp. 19-25

[3] Krinkin K, Filatov A, Filatov A, Huletski A, Kartashov D. Evaluation of modern laser based indoor slam algorithms. In: 2018 22nd Conference of Open Innovations Association (FRUCT). Helsinki, Finland. 2018. pp. 101-106

[4] Merzlyakov A, Macenski S. A comparison of modern general-purpose visual slam approaches. In: 2021 IEEE/RSJ International Conference on Intelligent Robots and Systems (IROS). Prague, Czech Republic. 2021. pp. 9190-9197

[5] Filatov A, Krinkin K. Multi-agent SLAM approaches for low-cost platforms. In: 2019 24th Conference of Open Innovations Association (FRUCT). Helsinki, Finland. 2019. pp. 89-95

[6] Thrun S, Liu Y. Multi-robot SLAM with sparse extended information filers. In: Robotics Research. The Eleventh International Symposium. Siena, Italy: Springer; 2005. pp. 254-266

[7] Kegeleirs M, Grisetti G, Birattari M. Swarm Slam: Challenges and perspectives. Frontiers in Robotics and AI. 2021;**8**:618268

[8] Gutmann J-S, Schlegel C. Amos: Comparison of scan matching approaches for self-localization in indoor environments. In: Proceedings of the First Euromicro Workshop on Advanced Mobile Robots (EUROBOT'96). Rome, Italy. 1996. pp. 61-67

[9] Fox D, Burgard W, Dellaert F, Thrun S. Monte Carlo localization: Efficient position estimation for Mobile robots. AAAI/IAAI. 1999;**1999**(343–349):2-2

[10] Yager RR. On the Dempster-Shafer framework and new combination rules. Information Sciences. 1987;**41**(2):93-137

[11] Raj T, Hashim FH, Huddin AB, Ibrahim MF, Hussain A. A survey on LiDAR scanning mechanisms. Electronics. 2020;**9**(5):741

[12] Blanc G, Mezouar Y, Martinet P. Indoor navigation of a wheeled mobile robot along visual routes. In: Proceedings of the 2005 IEEE International Conference on Robotics and Automation. Barcelona, Spain. 2005. pp. 3354-3359

[13] Paull L, Tani J, Ahn H, Alonso-Mora J, Carlone L, Cap M, et al. Duckietown: An open, inexpensive and flexible platform for autonomy education and research. In: 2017 IEEE International Conference on Robotics and Automation (ICRA). Marina Bay Sands, Singapore. 2017. pp. 1497–1504

[14] Benesty J, Chen J, Huang Y, Cohen I. Pearson correlation coefficient. In: Noise Reduction in Speech Processing. Berlin, Germany: Springer; 2009. pp. 1-4

[15] Myers L, Sirois MJ. Spearman correlation coefficients, differences between. Encyclopedia of Statistical Sciences. 2004;**12**:1-10

[16] Abdi H. The Kendall rank correlation coefficient. Encyclopedia of Measurement and Statistics. 2007;**2**: 508-510

[17] Croux C, Dehon C. Influence functions of the spearman and Kendall correlation measures. Statistical Methods & Applications. 2010;**19**(4):497-515

[18] Huletski A, Kartashov D, Krinkin K. Vinyslam: An indoor slam method for low-cost platforms based on the transferable belief model. In: 2017 IEEE/RSJ International Conference on Intelligent Robots and Systems (IROS). Vancouver, British Columbia, Canada. 2017. pp. 6770-6776

[19] Hess W, Kohler D, Rapp H, Andor D. Real-Time Loop Closure in 2d LIDAR SLAM. In: 2016 IEEE International Conference on Robotics and Automation (ICRA). Stockholm, Sweden. 2016. pp. 1271-1278

[20] Fallon M, Johannsson H, Kaess M, Leonard JJ. The Mit Stata Center dataset. The International Journal of Robotics Research. 2013;**32**(14):1695-1699

[21] Filatov A, Filatov A, Krinkin K, Chen B, Molodan D. 2d slam quality evaluation methods. In: 2017 21st Conference of Open Innovations Association (FRUCT). Helsinki, Finland. 2017. pp. 120-126

[22] Murphy K, Russell S. Rao-Blackwellised particle filtering for dynamic Bayesian networks. In: Sequential Monte Carlo Methods in Practice. Berlin, Germany: Springer; 2001. pp. 499-515

[23] Richardson M, Wallace S. Getting Started with Raspberry PI. Washington, D.C., USA: O'Reilly Media, Inc.; 2012

Section 4

Applications

Chapter 7

Aerial 3D Mapping with Continuous Time ICP for Urban Search and Rescue

Helge Andreas Lauterbach and Andreas Nüchter

Abstract

Fast reconnaissance is essential for strategic decisions during the immediate response phase of urban search and rescue missions. Nowadays, UAVs with their advantageous overview perspective are increasingly used for reconnaissance besides manual inspection of the scenario. However, data evaluation is often limited to visual inspection of images or video footage. We present our LiDAR-based aerial 3D mapping system, providing real-time maps of the environment. UAV-borne laser scans typically offer a reduced field of view. Moreover, UAV trajectories are more flexible and dynamic compared to those of ground vehicles, for which SLAM systems are often designed. We address these challenges by a two-step registration approach based on continuous time ICP. The experiments show that the resulting maps accurately represent the environment.

Keywords: UAV, mobile mapping, USAR, continuous time ICP, rescue robotics

1. Introduction

Reconnaissance is the basis for SAR missions and essential for strategic decisions. With the growing size of the site, it becomes difficult for officers in charge to get an overview of a disaster scenario, even in more frequent urban scenarios like houses on fire. Nowadays the main tools for scenario exploration used by firefighters are still 2D maps in command vehicles. Assessment is still done mainly by human visual inspection. Besides safety risks, this is a difficult and time-consuming process either for data collection and evaluation. Accordingly, this process aims the risk of missing important information.

In recent years, video footage in visible and thermal spectrum from small UAVs is increasingly used for scenario assessment and search operations [1]. UAV images are advantageous due to their overhead perspective and their capability to reach otherwise inaccessible areas. However, in practice, data evaluation is time-consuming likewise and faces the challenge of spatial correlation. Creating 3D maps in real-time is one useful approach to improve the evaluation process [2].

UAV-based approaches to 3D mapping often rely on structure from motion (SfM) techniques. Such approaches often do not provide 3D maps immediately, as dense mapping has high computational requirements [3]. Photogrammetric approaches are therefore mostly applied to large-scale scenarios like earthquakes [4, 5]. In [6], 3D point clouds generated from UAV images with SfM are used to localize and navigate unmanned ground vehicles. In large-scale scenarios, the immediate response phase has a time scale of hours to weeks [2], rather than minutes to hours as in more frequent urban scenarios, like houses on fire.

LiDAR-based mapping systems have the advantage of providing spatial information directly. Here the key issue is the implementation of an adequate SLAM approach. Unlike unmanned ground vehicles, UAVs are constantly in motion. Thus, the rigidity assumption does not hold true for sensors like laser scanners. To compensate for motion distortion in laser scans a continuous time formulation of the SLAM problem is convenient. [7] presents an application of continuous-time SLAM in UAV mapping. Groups of scans in a sliding window fashion are registered and the computed corrections are interpolated along the trajectory. In a second offline step, the trajectory is optimized globally. [8] use a map-centric approach. Instead of pose graph optimization, the map itself is deformed in case of loop closures.

LiDAR odometry and mapping (LOAM) [9] allows for real-time mapping by utilizing edge and planar features for registration. Distortion is removed by motion estimation from IMU mechanization and odometry estimation before registering to the map. An extension with visual odometry [10] was also demonstrated in aerial mapping [11]. Several approaches use feature extraction from LOAM and introduce environmental constraints such as ground planes [12] or tightly coupled LiDAR and INS [13]. They also incorporate loop closing, which LOAM does not. [14] propose improved edge and planar features. Due to offline batch optimization, the approach is not real-time capable. The study [15] instead relies on planes as landmarks and bundle adjustment for optimization. However, both methods are designed for indoor environments.

Instead of sampling at scan frequency in [16], the trajectory is modeled as a B-Spline to allow for interpolation between scan poses. The map is refined by realigning scans within local submaps. Based on the B-Spline trajectory representation, [17] registers multiple scans at once in a sliding window fashion to a sparse multiresolution surfel map, enabling real-time LiDAR odometry.

Thermal imaging is another important tool for firefighters [1]. It is used for navigation under low-sight conditions [18] (e.g., smoke) and to detect sources of fire and thermal hot spots [1]. Robotic systems equipped with thermal cameras also facilitate object detection [19] and search for casualties [20].

In a previous work [21], we gave an overview of our UAV system for SAR applications. Most similar to our setup is the UAV [22], integrated into a multirobot system with a focus on autonomous exploration [23]. In this paper, we present our mapping pipeline, generating accurate, temperature-enhanced 3D point clouds. The two-step registration approach is based on continuous time ICP.

2. Methodology

2.1 System overview

The proposed mapping system for rescue missions is divided into two principal components, an aerial segment for data acquisition and a ground segment for data processing and user interaction as depicted in **Figure 1**. The aerial segment is designed

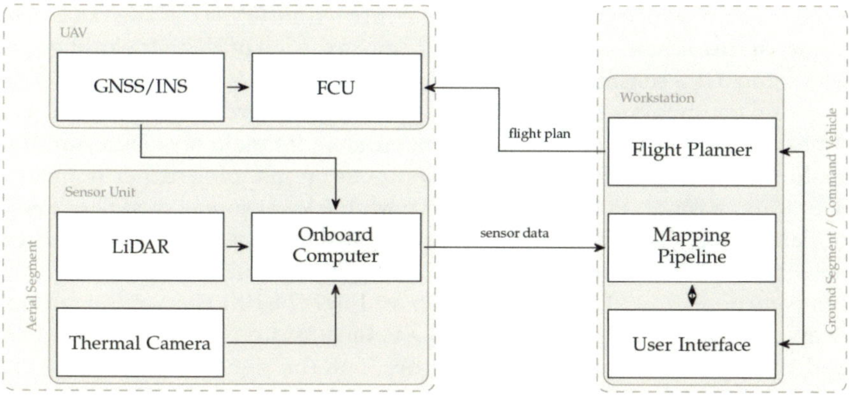

Figure 1.
System overview. The aerial segment is used for data acquisition while the ground segment runs the mapping pipeline and user interaction.

as a mapping sensor unit, that is able to operate without a UAV. The main sensor is a Velodyne Puck VLP16 Lite[1] laser scanner. With its low weight of 830 g and typical power consumption of 8 W, the sensor is appropriate for aerial mapping. It provides full 360° scans at a frequency of 10 Hz with a maximum range of 100 m. The vertical FoV (field of view) of 30° is sparsely covered by 16-line laser scans with an angular resolution of 2° vertically. The laser scanner is mounted at a pitch angle of 45°, resulting in a reduced usable horizontal FoV. This is a compromise between maximizing the ground coverage and ensuring a convenient overlap between consecutive scans for scan matching. As a second sensor, the system features an Optris PI 640LW[2] thermal camera to enrich the map with registered temperature information. In order to maximize the overlap with the laser scanner a wide-angle lens with an FoV of 90° x 64° is used. Thermal images are captured at a resolution of 640 px × 480 px with a frequency of 10 Hz. As an aerial platform, several Dji[3] UAVs are used that provide enough payload capacity to lift the sensor unit for a long period. The sensor unit mounted on the UAV is depicted in **Figure 2** on the left. For initial pose estimation, we

Figure 2.
UAV with mounted mapping payload in front of simulated facade fire (left) and the firefighter command vehicle (right).

[1] https://velodynelidar.com/products/puck-lite/
[2] https://www.optris.com/thermal-imager-optris-pi-640
[3] https://www.dji.com

rely on the trajectory reported by the flight control unit (FCU) of the UAV. This control system fuses three sets of redundant sensors, each of them including an IMU, a barometer and a GPS sensor.

The second component of our mapping system is the ground segment. Apart from communication for telemetry and data transmission, its main task is to run the mapping pipeline described in Section 2.2. As we focus on mapping rather than autonomous operation, a rough initial localization with UAV sensors is sufficient and onboard data processing is omitted. Furthermore, the ground segment offers online map visualization and a user interface for managing automated flights.

For this purpose, a workstation based on an Intel i7 8700 Hexacore running at 3.2 GHz and 16 GB RAM is integrated into a vehicle based on a german standard command vehicle (ELW1), depicted in **Figure 2** on the right.

2.2 Workflow

This subsection gives an overview of the processing pipeline, also visualized in **Figure 3**. Data acquired by the UAV is transmitted to the ground station. In the preprocessing stage, a pose for each laser scan is estimated by GNSS/INS localization. To reduce the computational effort, a keyframe approach is used. A new keyframe is selected only if the change in pose exceeds a defined threshold w.r.t. the previous keyframe, e.g., 1 m. We drop all scans that do not contribute to the map significantly,

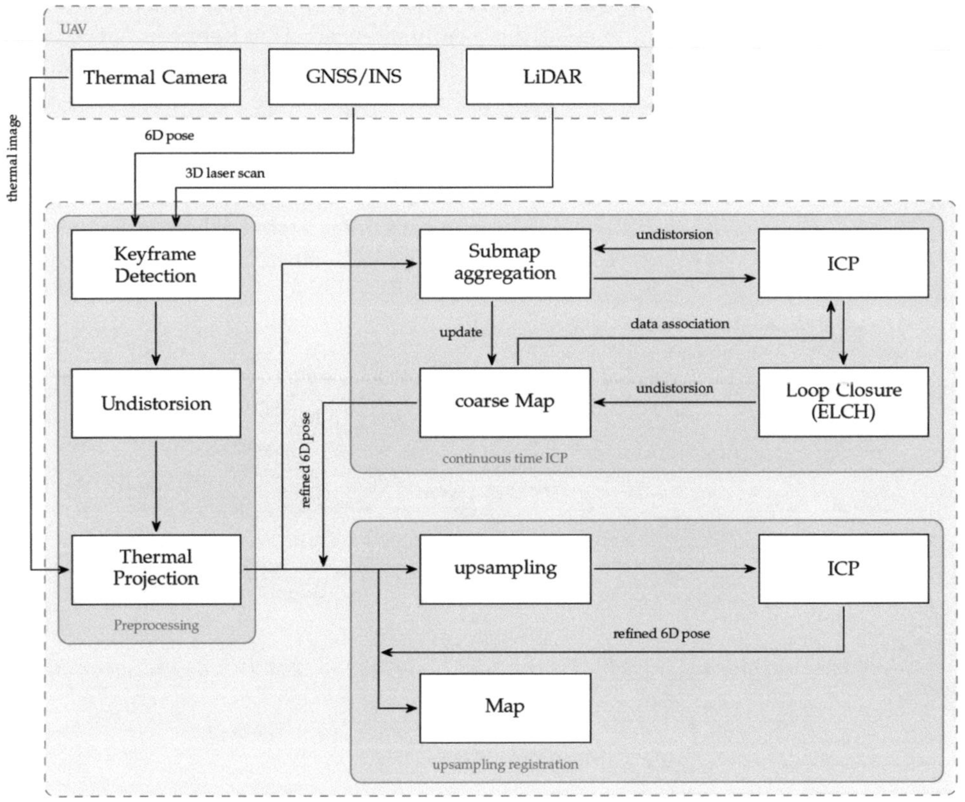

Figure 3.
Workflow of our UAV mapping system.

Figure 4.
Example for improved point clouds by proposed registration steps.

since for this application we are mainly interested in a map rather than navigation tasks. Selected frames are then undistorted based on the initial trajectory to compensate for motion. In an optional step, the laser scans are enriched with temperature information.

The mapping process is divided into two stages, for registration in a coarse to fine manner as shown in **Figure 4**. The core of the first registration stage is a continuous time ICP approach described in Section 2.4. This stage aims to reduce gross errors in the initial trajectory and to improve the map globally. Therefore a loop-closing technique adopted for the continuous time ICP is applied. This results in a coarse map of the environment, suitable for online inspection.

The second stage is run as a postprocessing step to further refine the trajectory and fine-register the laser scans. Here we use an upsampling approach to deal with the sparse nature of point clouds from a Velodyne VLP16 laser scanner. Details are given in Section 2.6.

The map is enriched with temperature information. Therefore, for each laser scan, we estimate the transformation between the poses of laser scanner and thermal camera at their current time stamps based on the GNSS/INS trajectory and project the thermal image on the laser scan. In our setup, this is done before matching the laser scans to enable fast thermal inspection of the coarse map. Note that by this approach only 3D points in the overlapping FoV of laser scanner and camera get temperature values assigned. Thus the temperature-enhanced map is less dense than the original map. In postprocessing, the mean temperature value of each voxel of the map is assigned to all points in the voxel. An example of the resulting representation is given in **Figure 5**. A facade fire scanned by the UAV is shown on the left, the corresponding temperature-enhanced 3D point cloud representation from our approach is shown on the right.

Figure 5.
UAV scanning a simulated facade fire on the campus of the Bavarian firefighter school Würzburg (left) and the corresponding 3D model colored by temperature (right). Red indicates high temperatures.

2.3 Continuous time SLAM

The movement of a mobile robot creates a trajectory $T = \{\mathbf{X}_0, \dots, \mathbf{X}_n\}$ where $\mathbf{X}_i$ is the 6-degree of freedom pose of the vehicle at time t_i. Using T a map $\mathbf{P} = \{\mathbf{p}_0, \dots, \mathbf{p}_n\}$ of the environment is derived, by transforming the set of timestamped measurements $M = \{\mathbf{m}_0, \dots, \mathbf{m}_n\}$ into the global coordinate system with $\mathbf{p}_i = \mathbf{X}_i \oplus \mathbf{m}_i = \mathbf{R}_i \cdot \mathbf{m}_i + \mathbf{t}_i$.

Given a sufficiently estimated trajectory, in [24], the map quality is improved by optimizing the trajectory, based on the ICP concept known for rigid registration. Similar to our graph-based SLAM algorithm [25], the n-scan registration problem is considered, but with a finer discretization of time, e.g., segments of a 3D scan or even single points. Under the assumption, that the error of the estimated trajectory is negligible for a short time interval the error function to be minimized is given by:

$$E_{i,j} = \sum_{k=i-N}^{i+N} \left\| \mathbf{X}_i \oplus \mathbf{m}_k - \mathbf{X}_j \oplus \mathbf{m}'_k \right\|^2 \tag{1}$$

where a small neighborhood of $2N$ points close in time to $\mathbf{m}_i$ the closest point $\mathbf{m}'_k \in M \setminus \{\mathbf{m}_{i-N}, \dots, \mathbf{m}_{i+N}\}$ is selected.

In other words, the measurements are grouped into overlapping submaps $\mathbf{M}_i = \{\mathbf{m}_{i-N}, \dots, \mathbf{m}_{i+N}\}$, where $\mathbf{m}_i$ denotes the reference measurement of the submap, that defines the corresponding pose $\mathbf{X}_i$. These submaps are then matched using the automatic high-precision registration of terrestrial 3D scans, i.e., the graph-based SLAM approach presented in [25, 26]. The graph is estimated using a heuristic that measures the overlap of the submaps based on the number of closest point pairs.

After applying globally consistent scan matching on the submaps the actual continuous time matching, as described in [24], is applied. Using the results of the rigid optimization as starting values for T, the numerical minimum of the underlying least square problem is computed.

2.4 Continuous time ICP

In contrast to the continuous-time SLAM approach, in ICP registration only the current pose is optimized with respect to its predecessor or a fixed map. Again consider the error of the initially given trajectory to be negligible in a small time interval before and after a pose $\mathbf{X}_i$. The positional error of $\mathbf{X}_i$ is then given by

$$E_i = \sum_{k=i-N}^{i+N} \left\| \mathbf{X}_i \oplus \mathbf{m}_k - \mathbf{p}'_k \right\|^2 \tag{2}$$

where $\mathbf{p}'_k$ is the closest point to m_k in $P \setminus \{\mathbf{p}_{i-N}, \dots, \mathbf{p}_n\}$. Note, that all poses $\mathbf{X}_j$ with $j < i - N$ are fixed.

For efficiency, we subsample the trajectory at $2N$ steps, as is done in the continuous time SLAM approach. The measurements $\mathbf{m}_k$ with $k = i \pm N$ then form a submap with pose $\mathbf{X}_i$. These submaps are then registered rigidly using plain ICP. Afterward the transformation update is locally distributed to all $2N$ poses between $\mathbf{X}_i$ and $\mathbf{X}_{i-1}$. Although not constrained, a linear distribution of translation and SLERP interpolation of rotation shows to be sufficient for small trajectory errors. For practical implementation, we form groups of 10 to 15 Velodyne VLP16 scans.

2.5 Loop closing

A common strategy in loop detection is, to rely on a heuristic based on pose-to-pose distance. This works well for trajectories with low drift, if either the flat surface assumption holds true, e.g., on ground vehicles and or if omnidirectional perception is available. In UAV laser scanning, both requirements are often not fulfilled. In our mapping system for instance the laser scanner is tilted by 45°, thus the effective FoV is reduced to less than 180° horizontally depending on the flight altitude and surrounding environment. As a consequence, there is often no overlap between loop closure candidates, or the overlap is small if the difference in orientation is high between the submaps. Instead, we search for poses with a maximum distance to the tilted plane, spanned by x and y coordinate of the sensor. Due to the reduced FoV, loop closure candidates additionally require an orientation that is similar to the query pose. After candidates are found, we apply the loop-closing technique ELCH from [27] to the submaps. To maintain the continuity of the trajectory we then distribute the transformation update of submaps to each laser scan in a similar way as in continuous time ICP in Section 2.4. An example is given in **Figure 6**.

2.6 Upsampling registration

Rigid registration of sparse point clouds, as those of a Velodyne VLP16, faces the problem of inhomogeneous point density in vertical and horizontal directions. The difference between the angular resolution between two scan lines and within one scan line is typically in the order of a magnitude. Similar to this, rotating 2D profilers provide 3D point clouds with a higher angular resolution within a sweep. This inhomogeneity in resolution has an impact on registration approaches using point-to-point distances as the ICP. As a result scan lines are pulled onto each other and the estimated transformation is distorted. In theory point to plane approaches perform better in such a case, as the underlying surface structure is estimated by local planar patches. However, the success of registration depends on the quality of the estimated normals

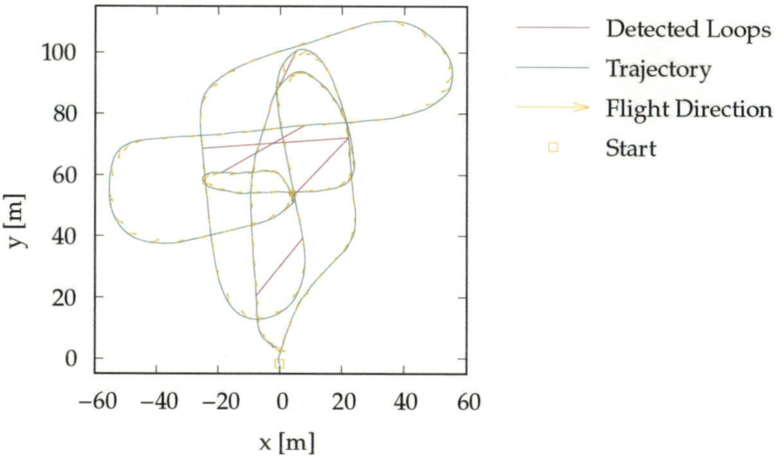

Figure 6.
Example for detected loop closures. We search for poses close to sensor xy plane. Poses with similar orientation are accepted as loop closing candidates.

of the points. The traditional approach of calculating the normals using k nearest neighbors is affected by the point distribution though.

To mitigate the problems caused by inhomogeneous point density, often assumptions about the underlying structure are made. One group of approaches relies on robust features, such as edges and planes [9]. The second group of approaches coarsely reconstructs the surface. [28] interpret a scan of a Velodyne 64 laser scanner as a range image and compute the linkage of the direct neighbors of a point in the image. Then they estimate the normal vectors as the average cross product of the vectors to the neighbors weighted by the linkage and apply a point-to-plane variant of ICP. A similar strategy is proposed by [29] that is more general in terms of the sensor model. Using the ordered structure of point clouds provided by sensors, they create a simple quad mesh to estimate the point normals and then apply a variant of the generalized-ICP [30]. Similar to this we approximate the surface by upsampling the point cloud. Inspired by the idea of range images, the sensor data is first organized into a ring and bin structure, preserving the real measurement. Then for each point, we create a line with its neighbor in the next ring and linear sample points. Compared to previous work [31], this simple approach proved to be sufficient due to the high point density within a ring. As in [29] an edge is rejected if it is nearly parallel to the line of sight with respect to the scan pose or a depth discontinuity is detected considering angular and range-dependent thresholds. The set of fitted line points forms a virtual scan that estimates the underlying surface, as visualized in **Figure 7**. Octree reduction generates a homogeneous distribution of points. The virtual scan is then rigidly registered against previous scans in their original form with ICP, either sequential or incremental, to further refine the robot trajectory.

2.7 Map organization

In both registration stages, we follow a scan-to-map matching strategy. A key issue in scan registration is the search for closest point pairs. The implementation of search trees in 3DTK [32] (e.g. octree) is optimized for fast point query operations and a

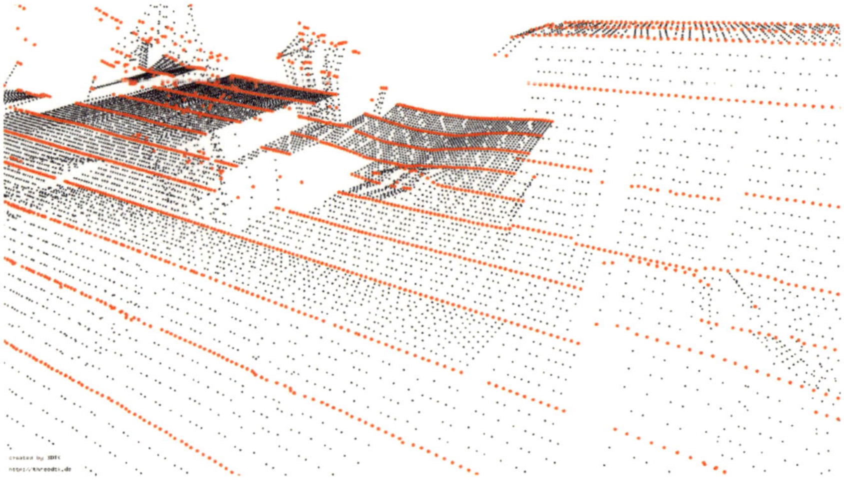

Figure 7.
Upsampled 3D laser scan from Velodyne VLP16. The original point cloud is depicted in red, the upsampled point cloud in black.

small memory footprint. However, due to their compact representation in memory, they do not allow for insert operation. Thus the search tree needs to be rebuilt once the map is updated. To deal with this, an efficient strategy for maintaining the map is important. For both registration steps, we consider the map to be a set of submaps. In the upsampling registration step, we follow a keyframe approach. Keyframes are added to the current active submap. Each keyframe in the active submap provides a search tree. All search trees of the submap are queried in parallel. Once a submap is finished, the keyframes are joined and a single search tree for the finished submap is built. A submap size of 10 to 15 scans has shown to be a good trade-off between increasing search time and reducing construction time for a new k-d tree containing the complete submap. To further optimize the runtime, we only consider a ROI (region of interest) during the registration. Therefore, we select the k nearest submaps each time starting a new submap. Each scan is then registered against this ROI map. As with the active submap, the search trees of all submaps are searched in parallel.

In the continuous-time ICP step, the map is organized in a similar fashion. The major difference is that the active submap is not composed of several search trees. Note that submaps do not change once they are completed unless a loop closure is detected. As a consequence, submaps may significantly overlap and thus cause redundancy. However, the continuity of the trajectory is maintained.

3. Results

To evaluate our approach we acquired a data set at the campus of the Bavarian Firefighter School Würzburg with our system described in Section 2.1. The UAV (DJI S1000) was manually flown in several loops at different altitudes above the site of the school. During the flight of 325 s a trajectory of 862 m was covered. We applied the pipeline as described in Section 2.2. The initial trajectory was provided by the UAV. We applied continuous time ICP with a maximum point-to-point distance of 0.75 m and a submap size of 10 VLP16 scans. The resulting point cloud and the flown trajectory are visualized in **Figure 8**.

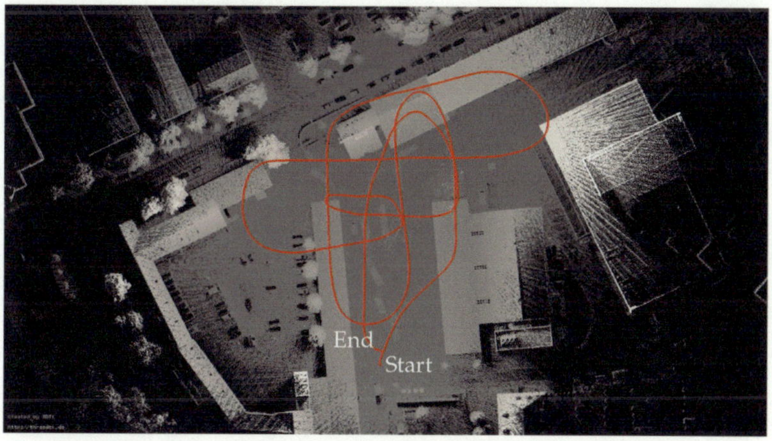

Figure 8.
Optimized point cloud from bird's eye view. The red line represents the UAV trajectory.

For comparison, we also applied BLAM! [33] and Pointmatcher [34] to the data set. Both approaches are based on ICP registration as our method is. To gain valuable results, we slightly modified BLAM, such that the UAV trajectory is used as an initial guess for odometry estimation.

For ground truth comparison we collected 42 highly precise laser scans with a Riegl VZ400 terrestrial laser scanner to cover the complete area of the firefighter school. The scans were registered using our ICP and SLAM methods [25] implemented in 3DTK [32] to obtain the reference point cloud.

As an error measure, we use the Absolute Trajectory Error (ATE) following [35]. It is computed as the root mean square error

$$\mathrm{ATE}_{\mathrm{pos}} = \sqrt{\frac{1}{N}\sum_{i=0}^{N-1} \|\Delta p_i\|^2}$$

$$\mathrm{ATE}_{\mathrm{rot}} = \sqrt{\frac{1}{N}\sum_{i=0}^{N-1} \|\angle(\Delta R_i)\|^2},$$

where Δp_i denotes the euclidean distance between the estimated and ground truth position at time step i of N and $\angle()$ denotes the rotation angle of the corresponding rotation ΔR_i in angle axis representation [35].

To obtain the ground truth trajectory we first localize each laser scan from the UAV system in the reference cloud. Then the initial trajectory as well as the optimized trajectories are aligned to the ground truth trajectory using all pose pairs and the ATE is calculated. The resulting mean errors are given in **Table 1**. **Figure 9** visualizes the positional error for all evaluated approaches and **Figure 10** the orientation error respectively.

Compared to the initial trajectory by GNSS/INS the mean error in position reduces to 0.172 m by applying our pipeline. Especially the drift in z-axis is removed as shown in **Figure 11**. Regarding orientation, the pitch angle θ_y provides the highest errors, as depicted in **Figure 12**. One explanation is that clocks of laser scanner and UAV are not synchronized. Thus the error increases during positive and negative acceleration and deceleration phases. θ_x and θ_z are less affected due to low speed and smooth curves in the trajectory. The mean rotational error is reduced 0.32°.

A comparison of the optimized point cloud to the reference point cloud justifies the results from ATE evaluation. Therefore, we computed the cloud-to-cloud distance with a maximum point-to-plane distance of 1 m. The results are visualized in **Figure 13** from two views. To support the visual representation, the corresponding error histogram is given in **Figure 14**. It shows that 90% of the points feature an error of less than 0.13 m. The accuracy of the Velodyne VLP16 is specified at ±3 cm, which corresponds with the peak of the histogram and is close to the median.

algorithm	$\mathrm{ATE}_{\mathrm{pos}}$ [m]	$\mathrm{ATE}_{\mathrm{rot}}$ [°]
GNSS	0.718	2.16
Ours	0.172	0.32
BLAM	0.383	1.74
Pointmatcher	0.225	0.32

Table 1.
Absolute trajectory error.

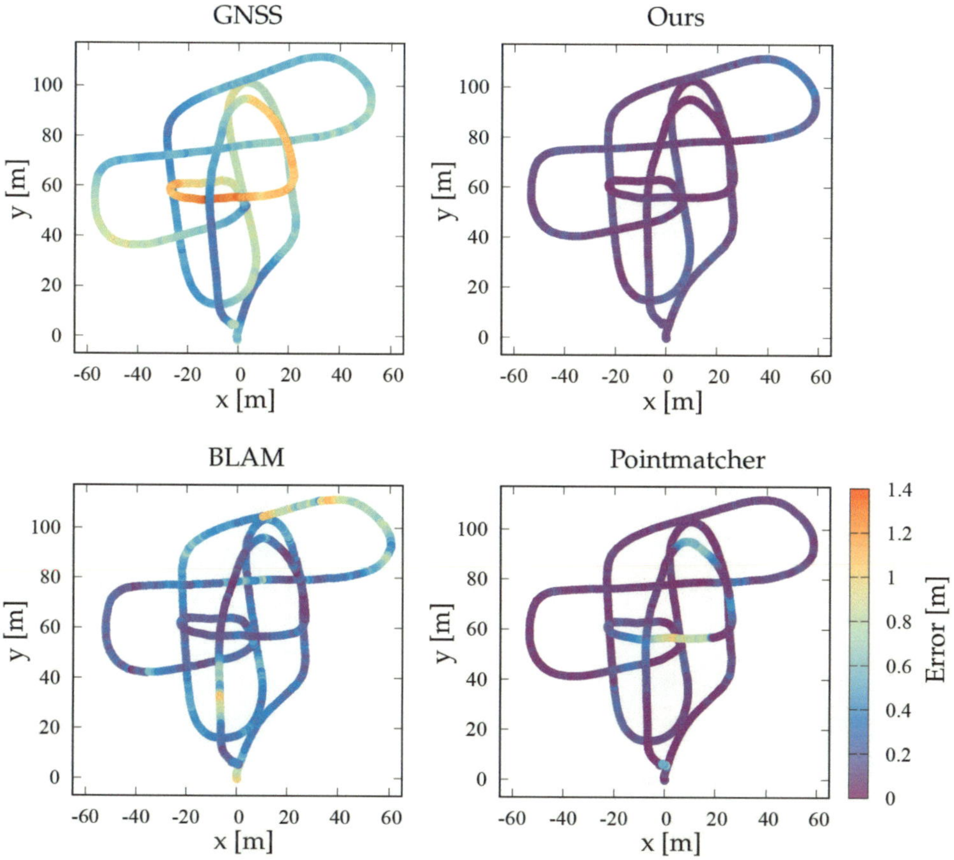

Figure 9.
Absolute trajectory error in position.

Note that **Figure 13** shows regions of high errors up to 1 m which do not reflect the low ATE errors discussed before. First, they refer to dynamic objects like cars that moved between the acquisition of the UAV data and the ground truth data (see **Figure 13**, at 1). Second, there are gaps in the ground truth data. Due to restricted possibilities to position the terrestrial laser scanner, it suffers from occlusions. This mainly affects the roofs of the buildings, e.g., no. 2 in **Figure 13**. Those gaps are filled by the airborne data set. As we used the distance to the closest plane as an error measure, some points at the borders are wrongly assigned a high error. A similar effect is present at windows and glass facades, e.g., at no. 4. Static areas, such as buildings and ground in majority, feature errors less than 0.1 m. Higher deviations up to 0.4 m, e.g., at no. 3, are explained by the fact, that this area was not directly overflown by the UAV, as depicted in **Figure 8**. Hence, the point density is reduced since the area was less often covered by the sensor and only from the side of the sensor at larger distances. Points in those areas have less influence on the scan matching than points in the core of the FoV since the probability to find corresponding points within the maximum search distance decreases. On the other hand, small registration errors have a greater impact on the distant areas of a scan. Moreover, the point density of the reference cloud is reduced in this area, biasing the error measure as described for number 2.

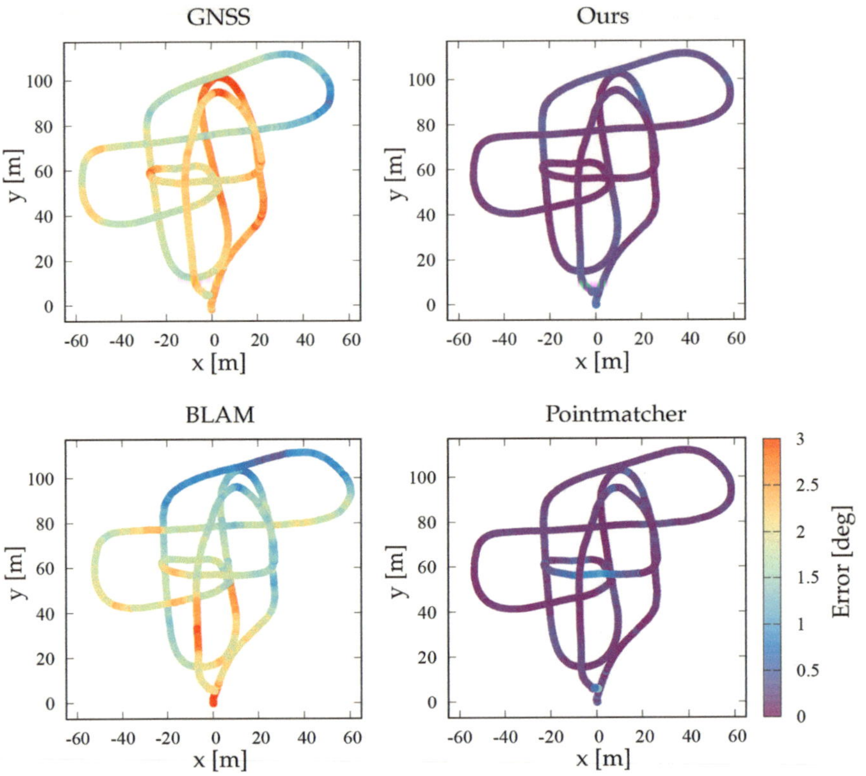

Figure 10.
Absolute trajectory error in orientation.

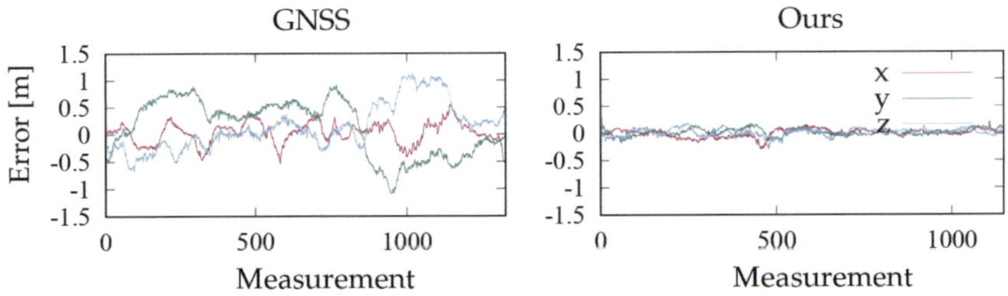

Figure 11.
Position error of individual axis. The error of GNSS trajectory (left) is reduced by the proposed pipeline (right).

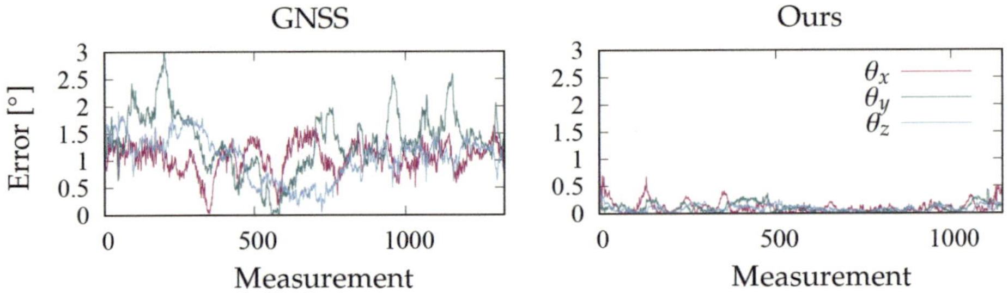

Figure 12.
Rotational error of individual axis. Error of GNSS trajectory (left) is reduced by the proposed pipeline (right).

To summarize the results, our approach is capable of generating detailed point clouds, accurately representing the environment.

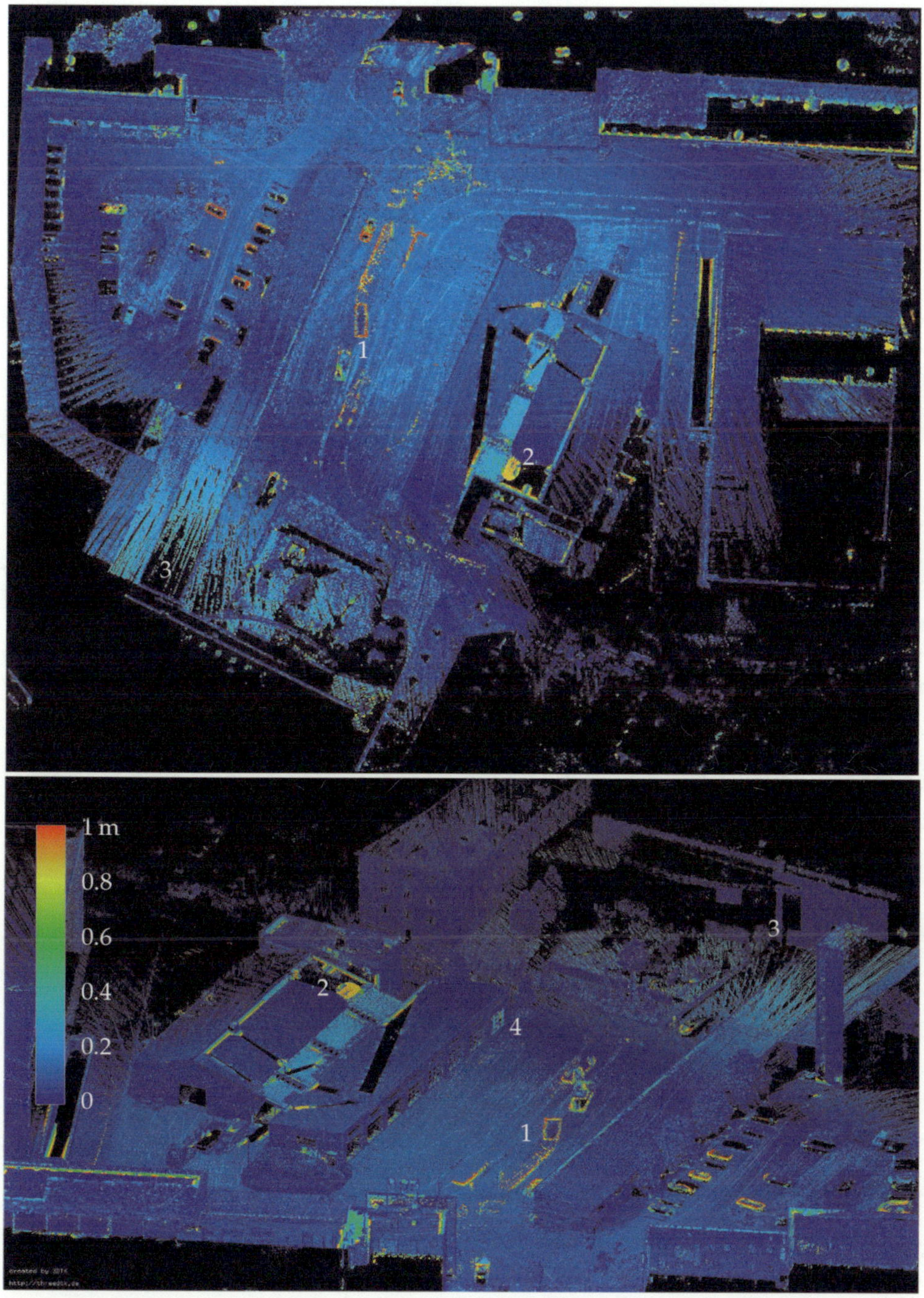

Figure 13.
Optimized point cloud from top (above) and oblique view (below). The color decodes the deviation from ground truth. Higher errors mainly correspond to dynamic objects (1) and occlusions in the reference data (2), as well as areas not directly overflown (3).

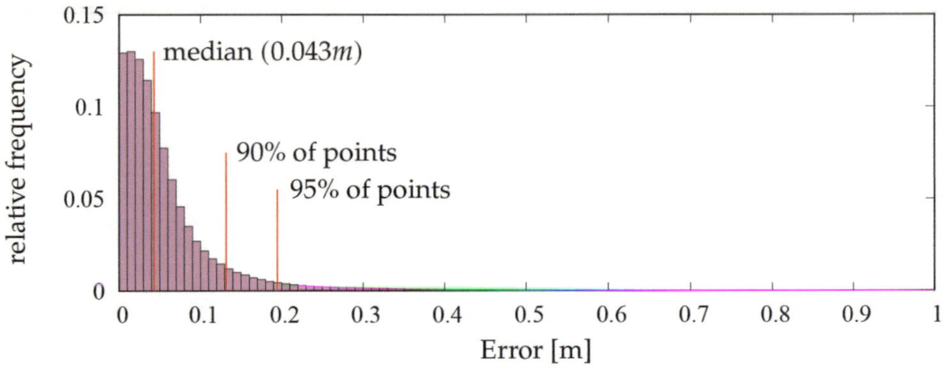

Figure 14.
Histogram of cloud-to-cloud error.

4. Conclusions

In this paper, we presented our pipeline for aerial 3D mapping in USAR scenarios. Using a two-step continuous time registration approach the system is able to produce an accurate representation of the environment. Enhanced with temperature values the generated 3D maps aim at facilitating the assessment of disaster scenarios. The current drawback of our approach is that only the first registration stage runs online, producing a coarse map while the second stage is designed as a postprocessing step producing an accurate map. Future work includes runtime optimization and further integration to enable online processing of the entire pipeline. Additionally point cloud analyzing methods, for instance, to detect heat sources are in work.

Acknowledgements

This work was funded by the projects Eins3D (FZK 13 N14183) and Deals3D (FKZ 13 N15313) from the Federal Ministry of Education and Research, Germany.

This publication was supported by the Open Access Publication Fund of the University of Würzburg.

Conflict of interest

The authors declare no conflict of interest.

Abbreviations

ATE	Absolute Trajectory Error
FoV	Field of View
GNSS	Global Navigation Satellite System
ICP	Iterative Closest Points
IMU	Inertial Measurement Unit
INS	Inertial Navigation System

LiDAR	Light Detection And Ranging
SAR	Search And Rescue
SLAM	Simultaneous Localization And Mapping
UAV	Unmanned Aerial Vehicle
USAR	Urban Search And Rescue

Author details

Helge Andreas Lauterbach*† and Andreas Nüchter†
Robotics and Telematics, Julius Maximilian University of Wüzburg, Würzburg, Germany

*Address all correspondence to: helge.lauterbach@uni-wuerzburg.de

† These authors contributed equally.

References

[1] Carrillo-Zapata D, Milner E, Hird J, Tzoumas G, Vardanega PJ, Sooriyabandara M, et al. Mutual shaping in swarm robotics: User studies in fire and rescue, storage organization, and bridge inspection. Frontiers in Robotics and AI. 2020;7:53

[2] Delmerico J, Mintchev S, Giusti A, Gromov B, Melo K, Horvat T, et al. The current state and future outlook of rescue robotics. Journal of Field Robotics. 2019;**36**(7):1171-1191

[3] Martell A, Lauterbach HA, Schilling K, Nüchter A. Benchmarking structure from motion algorithms of urban environments with applications to reconnaissance in search and rescue scenarios. In: Proceedings of the 16th IEEE International Symposium on Safety, Security, and Rescue Robotics (SSRR '18). Philadelphia, PA, USA: IEEE; 2018. pp. 1-7

[4] Verykokou S, Doulamis A, Athanasiou G, Ioannidis C, Amditis A. Uav-based 3d modelling of disaster scenes for urban search and rescue. In: 2016 IEEE International Conference on Imaging Systems and Techniques (IST). IEEE; 2016. pp. 106-111

[5] Kruijff-Korbayová I, Freda L, Gianni M, Ntouskos V, Hlaváč V, Kubelka V, et al. Deployment of ground and aerial robots in earthquake-struck amatrice in Italy (brief report). In: 2016 IEEE International Symposium on Safety, Security, and Rescue Robotics (SSRR). IEEE; 2016. pp. 278-279

[6] Gawel A, Dubé R, Surmann H, Nieto J, Siegwart R, Cadena C. 3d registration of aerial and ground robots for disaster response: An evaluation of features, descriptors, and transformation estimation. In: 2017 IEEE International Symposium on Safety, Security and Rescue Robotics (SSRR). Shanghai, China: IEEE; 2017. pp. 27-34

[7] Kaul L, Zlot R, Bosse M. Continuous-time three-dimensional mapping for micro aerial vehicles with a passively actuated rotating laser scanner. Journal of Field Robotics. 2016;**33**(1):103-132

[8] Park C, Moghadam P, Kim S, Elfes A, Fookes C, Sridharan S. Elastic lidar fusion: Dense map-centric continuous-time slam. In: 2018 IEEE International Conference on Robotics and Automation (ICRA). Brisbane, QLD, Australia: IEEE; 2018. pp. 1206-1213

[9] Zhang J and Singh S. Loam: Lidar odometry and mapping in real-time. In: Robotics: Science and Systems. Vol. 2. Berkeley, USA: Roboticsproceedings; 2014. p. 9. DOI: 10.15607/RSS.2014.X.007

[10] Zhang J, Singh S. Laser–visual–inertial odometry and mapping with high robustness and low drift. Journal of Field Robotics. 2018;**35**(8):1242-1264

[11] Zhang J, Singh S. Aerial and ground-based collaborative mapping: An experimental study. In: Field and Service Robotics. Springer International Publishing; 2018. pp. 397-412

[12] Shan T, Englot B. Lego-loam: Lightweight and ground-optimized lidar odometry and mapping on variable terrain. In: 2018 IEEE/RSJ International Conference on Intelligent Robots and Systems (IROS). Madrid, Spain: IEEE; 2018. pp. 4758-4765

[13] Shan T, Englot B, Meyers D, Wang W, Ratti C, Rus D. Lio-sam: Tightly-coupled lidar inertial odometry

via smoothing and mapping. In: 2020 IEEE/RSJ International Conference on Intelligent Robots and Systems (IROS)}. Las Vegas, NV, USA; IEEE; 2020. pp. 5135-5142. DOI: 10.1109/IROS45743.2020.9341176

[14] Gentil CL, Vidal-Calleja T, and Huang S. In2lama: Inertial lidar localisation and mapping. In: 2019 International Conference on Robotics and Automation (ICRA). Montreal, QC, Canada: IEEE: 2019. pp. 6388–6394

[15] Zhou L, Wang S, Kaess M. π-LSAM: LiDAR smoothing and mapping with planes. In: Proc. of IEEE Int. Conf. On Robotics and Automation (ICRA '21). Xi'an, China: IEEE; 2021. pp. 5751–5757. DOI: 10.1109/ICRA48506.2021.9561933

[16] Droeschel D, Behnke S. Efficient continuous-time slam for 3d lidar-based online mapping. In: 2018 IEEE International Conference on Robotics and Automation (ICRA). Brisbane, QLD, Australia: IEEE; 2018. pp. 1-9

[17] Quenzel J, Behnke S. Real-time multi-adaptive-resolution-surfel 6d lidar odometry using continuous-time trajectory optimization. In: 2021 IEEE/RSJ International Conference on Intelligent Robots and Systems (IROS). Prague, Czech Republic: IEEE; 2021. pp. 5499-5506

[18] S. P. Kleinschmidt SP and Wagner B. Visual multimodal odometry: Robust visual odometry in harsh environments. In: 2018 IEEE International Symposium on Safety, Security, and Rescue Robotics (SSRR). Philadelphia, PA, USA: IEEE; 2018. pp. 1–8

[19] Bañuls A, Mandow A, Vázquez-Martín R, Morales J, and García-Cerezo A. Object detection from thermal infrared and visible light cameras in search and rescue scenes. In: 2020 IEEE International Symposium on Safety, Security, and Rescue Robotics (SSRR). Abu Dhabi, United Arab Emirates: IEEE; 2020. pp. 380–386

[20] Feraru VA, Andersen RE, and Boukas E. Towards an autonomous uav-based system to assist search and rescue operations in man overboard incidents. In: 2020 IEEE International Symposium on Safety, Security, and Rescue Robotics (SSRR). Abu Dhabi, United Arab Emirates: IEEE; 2020. pp. 57–64

[21] Lauterbach HA, Koch CB, Hess R, Eck D, Schilling K, Nüchter A. The eins 3d project—Instantaneous uav-based 3d mapping for search and rescue applications. In: 2019 IEEE International Symposium on Safety, Security, and Rescue Robotics (SSRR). Würzburg, Germany: IEEE; 2019. pp. 1-6

[22] Kruijff-Korbayová I, Grafe R, Heidemann N, Berrang A, Hussung C, Willms C, et al. German rescue robotics center (drz): A holistic approach for robotic systems assisting in emergency response. In: 2021 IEEE International Symposium on Safety, Security, and Rescue Robotics (SSRR). New York City, NY, USA: IEEE; 2021. pp. 138-145

[23] Schleich D, Beul M, Quenzel J, and Behnke S. Autonomous flight in unknown gnss-denied environments for disaster examination. In: 2021 International Conference on Unmanned Aircraft Systems (ICUAS). Athens, Greece: IEEE; 2021. pp. 950–957. DOI: 10.1109/ICUAS51884.2021.9476790

[24] Elseberg J, Borrmann D, Nüchter A. Algorithmic solutions for computing accurate maximum likelihood 3D point clouds from mobile laser scanning platforms. Remote Sensing. MDPI; 2013; 5:5871-5906

[25] Borrmann D, Elseberg J, Lingemann K, Nüchter A, Hertzberg J.

Globally consistent 3d mapping with scan matching. Journal Robotics and Autonomous Systems (JRAS). Elsevier; 2008;**56**:130-142

[26] Borrmann D, Elseberg J, Lingemann K, Nüchter A, Hertzberg J. The efficient extension of globally consistent scan matching to 6 DoF. In: Proceedings of the 4th International Symposium on 3D Data Processing, Visualization and Transmission (3DPVT '08). Atlanta, USA: Georgia Institute of Technology; 2008. pp. 29-36

[27] Sprickerhof J, Nüchter A, Lingemann K, Hertzberg J. An explicit loop closing technique for 6D SLAM. In: Proceedings of the 4th European Conference on Mobile Robots (ECMR '09). Mlini/Dubrovnic, Croatia: KoREMA; September 2009. pp. 229-234

[28] Moosmann F, Stiller C. Velodyne slam. In: 2011 IEEE Intelligent Vehicles Symposium (IV). Baden-Baden, Germany: IEEE; 2011. pp. 393-398

[29] Holz D and Behnke S. Registration of non-uniform density 3d point clouds using approximate surface reconstruction. In: Proc. of the 41st Int. Symposium on Robotics (ISR) and 8th German Conference on Robotics (ROBOTIK). Munich, Germany: VDE; 2014. pp. 1-7

[30] Segal A, Hähnel D, and Thrun S. Generalized-icp. In: Proceedings of Robotics: Science and Systems. Seattle, WA, USA: The {MIT} Press; 2009. pp. 161-168. DOI: 10.15607/RSS.2009.V.021

[31] Lauterbach HA, Borrmann D, Nüchter A, Rossi AP, Unnithan V, Torrese P, et al. Mobile mapping of the la corona lavatube on lanzarote. In: Proceedings of the ISPRS Geospatial Week 2019, Laserscanning 2019, ISPRS Annals Photogrammetry and Remote Sensing, Spatial Inf. Sci., IV-2/W5. Enschede, Netherlands: Copernicus Publications; 2019. pp. 381-387

[32] Nüchter A. 3DTK - the 3D toolkit. 2020 https://www.threedtk.de. [Accessed: March 8, 2021]

[33] Nelson E. BLAM! – Berkeley Localization and Mapping. 2016 https://github.com/erik-nelson/blam. [Accessed: March 23, 2021

[34] Pomerleau F, Colas F, Siegwart R, Magnenat S. "Comparing icp variants on real-world data sets." Autonomous Robots. Springer International Publishing; April 2013;**34**:133-148

[35] Zhang Z, Scaramuzza D. A tutorial on quantitative trajectory evaluation for visual (−inertial) odometry. In: 2018 IEEE/RSJ International Conference on Intelligent Robots and Systems (IROS). Madrid, Spain: IEEE; 2018. pp. 7244-7251

Chapter 8

Practical Insights on Automotive SLAM in Urban Environments

Piotr Skrzypczyński

Abstract

This chapter tackles the issues of simultaneous localization and mapping (SLAM) using laser scanners or vision as a viable alternative to the accurate modes of satellite-based localization, which are popular and easy to implement with modern technology but might fail in many urban scenarios. This chapter considers two state-of-the-art localization algorithms, LOAM and ORB-SLAM3 that use the optimization-based formulation of SLAM and utilize laser and vision sensing, respectively. The focus is on the practical aspects of localization and the accuracy of the obtained trajectories. It contributes to a series of experiments conducted using an electric car equipped with a carefully calibrated multisensory setup with a 3D laser scanner, camera, and a smartphone for collecting the exteroceptive measurements. Results of applying the two different SLAM algorithms to the data sequences collected with the vehicle-based multisensory setup clearly demonstrate that not only the expensive laser sensors but also monocular vision, including the commodity smartphone camera, can be used to obtain off-line reasonably accurate vehicle trajectories in an urban environment.

Keywords: SLAM, LiDAR, vision, GNSS, factor graph, evaluation

1. Introduction

In the last decade, we have witnessed a rapid development of methods, algorithms, and technologies that make vehicles more autonomous. This trend focuses on self-driving cars but includes also public transportation vehicles and advanced driver assistance systems. In all cases, the knowledge of the vehicle's pose and the availability of an internal world model are enabling conditions for efficient task and motion planning, reasoning about the environment's semantics, and man–machine interaction.

Although the most used localization solution in automotive applications is the global navigation satellite system (GNSS), the availability of GNSS signal is limited in many practical scenarios, such as driving through a tunnel, maneuvering in an underground parking lot, or navigating among tall buildings in the downtown. Therefore, simultaneous localization and mapping [1] and visual odometry (VO) [2] are considered a complementary means of yielding vehicle pose estimates that enable safe navigation of autonomous or semi-autonomous vehicles. While there is a large body of research on SLAM and VO, there are still many open issues when it comes to more complicated yet more practical cases, including 3D perception and lifelong map learning [3].

The two sensing modalities that dominate automotive SLAM are passive cameras and 3D laser scanners, known as LiDARs (light detection and ranging). Both classes of sensors have important advantages, but also some limitations. Active 3D laser sensors are well-suited to work under any lighting conditions, while passive cameras struggle with poorly illuminated scenes and sudden lighting changes. LiDAR sensors provide reasonably dense depth images within the range of tens, or even hundreds of meters. These data make it possible to build dense maps of the environment, while passive vision SLAM systems usually rely on sparse maps of point features. However, passive cameras are certainly the most affordable and easy-to-use sensors that can be deployed at minimal cost, while visual SLAM systems can achieve satisfying accuracy of trajectory estimation [4].

This chapter presents an experimental study on trajectory estimation accuracy by two popular open-source SLAM systems in the context of navigation for autonomous vehicles in urban environment. It contributes a unique set of experiments conducted using an autonomous electric vehicle equipped with a rich set of exteroceptive sensors. An important part of these experiments is the ground truth trajectories collected using an accurate RTK-GPS (real-time kinematics global positioning system). These data allow us to test two state-of-the-art SLAM solutions: LiDAR odometry and mapping (LOAM) [5] and ORB-SLAM3 [4] using two sensing modalities: LiDAR and passive vision, respectively.

Additionally, while performing these experiments we collected also images from a camera of a commodity smartphone attached to the windshield of the vehicle. This low-cost camera serves as a test bed for minimal-cost localization in urban environments, as it can be attached to any vehicle.

The aim of the presented research is to answer the question of how accurate are the trajectories obtained using different SLAM solutions and from different sensing modalities in scenarios imitating urban driving but under full control of an accurate GNSS solution providing precise ground truth trajectories.

The collected data are processed off-line, as often SLAM is used to obtain reference positions for data collected by a vehicle, which is the case of the CityBrands project, the presented research is part of, where trajectories are collected to identify locations of billboards and other advertisement installations. One of the purposes of this project is to verify the thesis that it is possible to obtain vehicle pose estimates of a local translational error not exceeding 1 m using only a smartphone's camera as a sensor.

2. Concept and implementation of experiments

The experiments were carried out using a set of integrated sensors attached rigidly to a common frame and mounted on a vehicle. A Melex electric vehicle (work car) was used during the experiments. The rigid frame with the sensors was mounted on its roof, and data was collected by a computer mounted in the luggage compartment, except for data from a smartphone, stored directly in its memory. The basic component is an aluminum frame with a GNSS receiver and an Ouster OS1-128 Gen2 3D laser scanner with a range of 150 m and a 360° horizontal field of view. Other components of the system: a Basler acA1440-220um USB3 camera with a Basler C125-0418-5M F1.8 f4mm lens and an AHRS (attitude and heading reference system) Xsens MTi-30-2A8G4 sensor (**Figure 1A**).

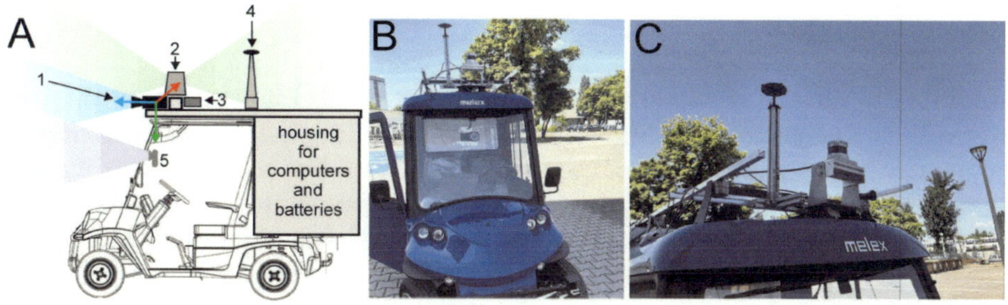

Figure 1.
Melex electric vehicle with a sensory system configured for the experiments: 1 – Camera, 2 – LiDAR, 3 – IMU, 4 – GNSS, 5 – Smartphone (A), a smartphone visible behind the vehicle's windshield (B), and a close-up of the main sensors on the roof (C).

The whole sensor set was calibrated both in terms of internal camera parameters (intrinsic) and in terms of external parameters (extrinsic) defining transformations between sensor coordinate systems. All these onboard exteroceptive sensors were integrated under Linux using robot operating system (ROS) [6] with ROS nodes deployed for each of the sensors, writing to defined ROS topics, and finally, saving the data using the flexible "rosbag" format. The ROS system provided also time synchronization to all these nodes by proper time stamping of the messages written to ROS topics and then to rosbag files.

Including in the multisensory system, a Samsung Galaxy A20 smartphone (**Figure 1B**) used as a second camera requires mutual calibration of the external parameters (translation and rotation) of the smartphone's camera relative to the Basler camera (**Figure 1C**). Since the smartphone runs Android, it is not connected to the system as a ROS node and is not subject to system time synchronization. The images from the smartphone's camera were collected using a dedicated Android application, as the default Samsung app that supports the camera on the A20 phone does not allow setting a specific focus value (locking autofocus). With automatic settings of the camera's focus and white balance, it was not possible to calibrate it accurately.

2.1 Calibration

Calibration is essential in a multi-sensor system that is dedicated to provide data for comparison of different methods and sensing modalities. In our case, calibration was performed offline, on data sequences that were collected before starting the tests, applying a number of different tools.

The main software package used to calibrate the sensors was Kalibr [7], a system for calibration of multi-camera and camera-IMU (inertial measurements unit) systems. Kalibr is integrated with ROS, which made many operations on data stored in rosbag files much easier. The entire calibration process was accomplished in several steps:

1. Calibration of the Basler camera parameters using Kalibr.

2. Calibration of the Basler camera with the Xsens AHRS (which is a type of IMU), implemented with Kalibr and the frame with sensors removed from the vehicle,

as the procedure required to move the camera/IMU system in order to produce appropriate data.

3. Calibration of the Basler camera with the Ouster OS1 LiDAR, which was conducted using the novel spatiotemporal calibration method [8] that makes it possible to obtain accurate translation and rotation between the coordinate systems of both sensors, but provides also an estimate of the time difference between the image frames and laser scans. Note that in our setup there is no hardware trigger for the sensors, and the time is only synchronized by ROS time stamps, which requires estimating the actual difference imposed by hardware.

4. Calibration of the parameters of the Samsung Galaxy A20 camera.

5. Calibration of the external parameters between the Basler camera and the smartphone's camera, considered as a stereo pair.

Surprisingly, the last two tasks turned out to be most difficult. An attempt to use Kalibr for these calibrations failed, probably because the smartphone's camera has a rolling-shutter frame, and the Kalibr procedure was unable to find some of the corners of the checkerboard pattern used for calibration (**Figure 2A**). The resulting camera parameters, although looking correct, were inappropriate for the ORB-SLAM3 system, causing problems with the initialization of the system (**Figure 2B** and **C**). Finally, the Galaxy A20 camera was calibrated using the classic approach [9] and software that allowed us to manually point the calibration corners on the checkerboard. These parameters allow ORB-SLAM3 to initialize correctly (**Figure 2D**). The resulting calibration parameters of the relative translations and rotations of the sensors were used to build a TF tree in the ROS environment (**Figure 2E**).

2.2 Ground truth acquisition

The solution for obtaining the reference (ground truth) trajectory is a satellite navigation system. In order to achieve the centimeter-level accuracy required in the

Figure 2.
Calibration of the multisensory system on the vehicle: Calibration checkerboard shown to the galaxy A20 smartphone's camera (A), initialization of the ORB-SLAM3 system on an image from galaxy camera (B), ORB-SLAM3 incorrectly initialized because of improper calibration of the camera (C), point features tracked correctly by ORB-SLAM3 when the camera is correctly initialized (D), at the ROS RViz view of the transformations tree in the fully calibrated multisensory system (E).

application under consideration, it is necessary to transmit to the RTK-GPS system corrections coming from a ground station with a known geographic position [10]. The setup used in the experiments applied an Ublox C099-F9P sensor with a ZED-F9P module for RTK-GPS, which can provide a localization accuracy of 2 cm with a position frame rate of 10 Hz when operating in L1/L2 modes. With access to commercial RTK corrections transmitted via the Internet and LTE modem, it was not necessary to establish a separate reference station at a known position.

3. SLAM problem

Over the past few years, traditional SLAM algorithms based on filtering (extended Kalman filter, particle filter) [11] have been replaced by methods based on optimizing graph representations of the SLAM problem [12], most often in the form of a sensor poses graph [13] or factor graph [14] binding multiple variables subject to optimization, describing both the state of the agent (vehicle/sensor) and a map of the environment in the form of features.

As shown in [12], the SLAM problem can be viewed in terms of factor graph optimization. The factor graph formulation can be applied to many forms of SLAM and localization, regardless of the used sensing modality (e.g., to WiFi fingerprints [15]), and regardless of the used features/landmarks, for example, visual point features [4, 16] or LiDAR-based segment features [17]. The most general form, a method known in the computer vision field as structure from motion (SfM), assumes that all historical sensor positions are related to observed features through measurements. These relationships can be thought of as a random Markov field with variables x describing the sensor poses and variables f describing the features (**Figure 3A**). However, if SLAM is used not for a limited scene, but a large area, the factor graph grows when new positions are added when the vehicle moves, while new features appear when the vehicle/sensor explores new parts of the environment.

Processing a very large factor graph should be avoided due to computational performance limitations. For this reason, pose-based graph SLAM systems marginalize all historical features (and thus do not have an explicit map) and keep only the pose graph with constraints between them (**Figure 3B**). These constraints are derived from the estimated movement of the vehicle and are formed locally, within a certain window along the trajectory, solving a problem analogous to visual odometry. Constraints (factor graph arcs) can also represent loop closures, that is, relationships between locations that are distant in the position chain but lie close enough spatially to be observed simultaneously. Constraints on loop closures are essential to the

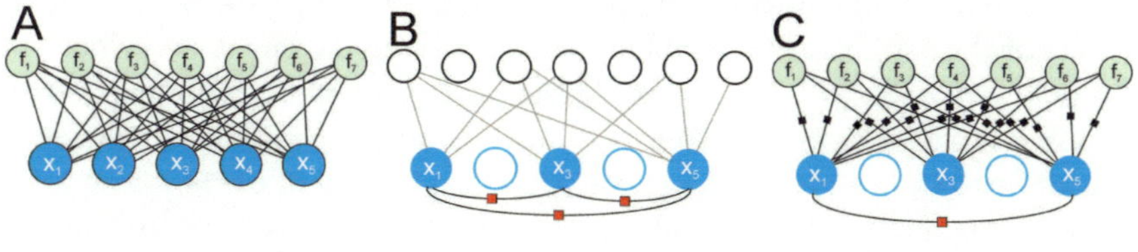

Figure 3.
Markov random field for the SfM problem (A). Factor graph for position-only SLAM (B) and factor graph for bundle adjustment SLAM (C). Larger blue circles are vehicle/sensor positions, green circles are features considered in optimization, empty (white) circles are those features and poses that have been marginalized, graph arcs are measurements, and rectangles denote factors from pose-to-pose constraints (red) or pose-to-feature constraints (black).

successful implementation of the SLAM task, as they allow the system to reduce trajectory drift compared to a comparable visual odometry system. The constraints are represented on the graph as factors, shown in **Figure 3B** by small rectangles. In the pose graph formulation of SLAM [13] direct observations of the features are not used, as these measurements are utilized only to compute the constraints between agent poses in the graph. In pose graph optimization new pose values are estimated that minimize the following error:

$$\underset{\mathbf{T}_1,\dots,\mathbf{T}_n}{\operatorname{argmin}}\left(\sum_i h(\mathbf{T}_i,\ \mathbf{T}_{i+1})\Omega_{i,i+1}h(\mathbf{T}_i,\ \mathbf{T}_{i+1}) + \sum_j h\left(\mathbf{T}_{l(j,\ 1)},\ \mathbf{T}_{l(j,\ 2)}\right)\Omega_{l(j)}h\left(\mathbf{T}_{l(j,\ 1)},\ \mathbf{T}_{l(j,\ 2)}\right)\right), \tag{1}$$

where $\mathbf{T}_1,\dots,\mathbf{T}_n$ are the optimized sensor poses, $h(\mathbf{T}_i,\ \mathbf{T}_{i+1})$ is the pose error between the i-th and $(i+1)$-th poses from the measurement of the visual odometry process, and $h\left(\mathbf{T}_{l(j,\ 1)},\ \mathbf{T}_{l(j,\ 2)}\right)$ is the pose error of the j-th loop closure measurement defined as the error between $l(j,\ 1)$-th and $l(j,\ 2)$-th poses related by this loop closing event.

In contrast, a full SLAM system uses a nonlinear estimation technique [18] to optimize the constraint graph, which, however, has a much larger number of factors directly linking feature positions to sensor poses (**Figure 3C**). Also not all poses are used directly—only those considered as keyframes are stored in the map. The choice of keyframes strongly depends on the used sensing modality and the processing of features but is crucial to both the efficiency and robustness of SLAM [19]. An edge represents the resulting constraint from a sensor measurement between the i-th position and j-th feature. The uncertainty of each constraint is represented by its information matrix Ω, which can be determined by inverting the covariance matrix of a particular measurement [19]. This approach is similar to the bundle adjustment (BA) method used to efficiently solve the SfM problem [20] but can be applied in real time and used for large optimization problems (thus large maps) due to its computational efficiency [21]. The optimization step can be described mathematically as:

$$\underset{\mathcal{T},\mathcal{F}}{\operatorname{argmin}}\left(\sum_{i=1}^{n}\sum_{j\in\mathcal{F}_i} f\left(\mathbf{f}_i,\ \mathbf{T}_\mathbf{j}\right)\Omega_{i,j}f\left(\mathbf{f}_i,\ \mathbf{T}_\mathbf{j}\right) + \sum_k h\left(\mathbf{T}_{l(k,\ 1)},\ \mathbf{T}_{l(k,\ 2)}\right)\Omega_{l(k)}h\left(\mathbf{T}_{l(k,\ 1)},\ \mathbf{T}_{l(k,\ 2)}\right)\right), \tag{2}$$

where $\mathcal{T}$ is a set of all optimized sensor poses, $\mathcal{F}$ is the set of all optimized features, $\mathcal{F}_i$ defines the sets of indices of all features visible from i-th sensor pose, and $f\left(\mathbf{f}_i,\ \mathbf{T}_\mathbf{j}\right)$ is the error function of the measurement between the i-th feature and j-th pose. Note that the part of the optimization problem related to loop closing is described the same way as in (1), with $h\left(\mathbf{T}_{l(k,\ 1)},\ \mathbf{T}_{l(k,\ 2)}\right)$ being the pose error of the k-th loop closure measurement. The SLAM formulations (1) and (2) take the notion of topological loop closing, as defined in [19], which requires detecting the event of revisiting an already mapped area and constructing an additional pose-to-pose constraint by matching the reobserved features to the known part of the map. However, [19] points out that there is also geometric loop closing possible, where the vehicle/sensor does not need to detect the revisit event but matches the observed features to the known map using local criteria. These local criteria differ depending on the type of features. In LiDAR-based SLAM with planar or linear features, geometric distances are applied to discriminate incorrect matches [22], while in visual SLAM with point features [16], robust matching of descriptors can be employed [23]. Anyway, this geometric loop closing is only local, because if a very large loop has to be closed the accumulated pose drift may prevent our agent from executing the correct matching of the local features to the map. This is not the case for the loop closing formulation shown in (1) and (2), because in topological loop closing the respective locations (poses) are matched using global, appearance-based methods [24]. Note that the conceptual diagram in **Figure 3C** shows both ways of loop closing on a single graph.

3.1 LOAM system

The open-source LiDAR odometry and mapping is a state-of-the-art solution that ranked first in car localization based on the KITTI benchmark [25]. The processing of laser scans in LOAM is divided into several stages for feature extraction, odometry, mapping, and integration. The system operates on detected planar and edge features, which make it possible to reduce the extensive 3D laser scanner data stream. These features are then matched between the current and previous scans. In this step, the LiDAR sensor pose (and indirectly the vehicle pose) is estimated using constraints set by the associated features. The LOAM system assumes that scanning takes place continuously while the scanner is in motion. Therefore, after the odometry stage, the measured points are projected onto the coordinate system associated with the starting point of the scanning process, and a geometrically corrected point cloud is obtained. This corrected cloud is the input to the mapping procedure, which optimizes the match between the new cloud from the sensor and the global map (**Figure 4A**), giving a more accurate pose estimate than that obtained in the odometry step. Map optimization is performed less frequently (1 Hz) than odometry (10 Hz) because this step is more computationally intensive. The next localization step combines the pose from the map optimization with the latest available pose from laser odometry to provide the best available sensor pose estimate at the time. Note that LOAM does not use explicit (i.e., topological) loop closing. Although it is possible to implement this type of loop closing in a LiDAR-based SLAM with planar and line features [17], LOAM uses an approach that can be rather seen as instantaneous, geometric loop closing by matching the local features to the already known part of the map.

The LOAM system is set up to work with the Velodyne VLP-16 scanner; however, the scans from the Ouster OS1-128 sensors were used in the described study, modifying the program code accordingly. The modifications made included parameters

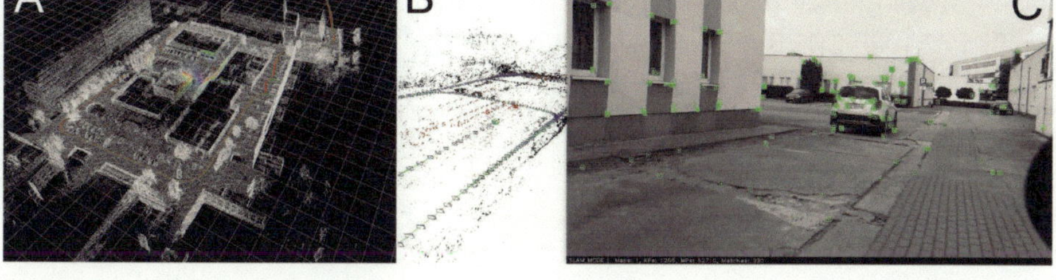

Figure 4.
Visualization of the LOAM global point cloud map during one of the experiments (A), the point features extracted by ORB-SLAM3 in its internal map representation (B), and detected directly on an image (C).

related to different horizontal and vertical observation angles, as well as necessary changes in assumptions about how points are acquired in each scan to correctly represent measurements at the beginning of the scan based on the current motion estimate.

3.2 ORB-SLAM3 system

The visual SLAM system ORB-SLAM3 is a variant of the ORB-SLAM algorithm presented in Ref. [16]. It was developed as a monocular SLAM using the concept of point feature map and bundle adjustment optimization. The third version, available as open source software [4], can use data from a single camera (monocular SLAM), stereo camera, or an RGB-D camera providing depth measurements. Constraint graph optimization is implemented in ORB-SLAM3 using the g^2o library [18]. The system architecture is divided into a front-end and a back-end storing the map (**Figure 4B**). ORB-SLAM3 retains all the features of monocular SLAM, also taking into account features devoid of depth information and optimizing the map only on the basis of feature reprojection errors in the images. In addition, ORB-SLAM3 fully implements loop closure based on scene view recognition, using DBoW2 bag-of-visual-words algorithm [24]. This makes it possible to efficiently close loops of any size, not just local loops. Note that although the detection of loop closures in ORB-SLAM3 is appearance-based, the optimization in factor graph after closing a loop is accomplished on the level of features rather than pose-to-pose constraints. In ORB-SLAM3, the fully multi-scale ORB features [26] are used throughout the system: for tracking sensor motion (**Figure 4C**), matching the current perception to the map (creating constraints), and closing loops. ORB-SLAM3 uses optimization with the factor graph concept at several stages of processing, not only for global optimization, as shown in (2), but frequently performs local bundle adjustment over several keyframes and their associated features to obtain local map reconstruction. A robust Huber kernel [27] can be used for optimization with reprojection error to reduce the influence of spurious associations.

During tracking, a simple model of camera motion is used, assuming motion at a constant speed. This model can be considered sufficient in light of the scenarios of the experiments conducted, in which the Melex vehicle moved at speeds from 10 to 30 km/h. Although ORB-SLAM3 can work also as visual-inertial SLAM, which makes it much more robust, for the sake of comparison between results from the two cameras used in the presented setup the application of ORB-SLAM3 was restricted to monocular SLAM only.

A more challenging aspect of applying ORB-SLAM3 in the conducted experiments is the use of a smartphone as a camera. The developers of ORB-SLAM3 do not provide a separate implementation for mobile devices and do not support Android. However, as offline processing of the collected data is assumed, only the quality of the images and the availability of correct calibration data are important, because the images from Samsung A20 are further processed on an x86 computer under Linux.

4. Methodology for assessing the accuracy of SLAM

With regard to the VO and SLAM algorithms, it is possible to talk about the accuracy of estimation of two components of the created model: the sensor trajectory and the map itself. Both of these components are of practical importance in the tasks of navigation and recognition of the environment, and they are interrelated since the accuracy of the trajectory estimation determines the accuracy of the registration of the position of the features that form a map of the environment.

A standard sensor for providing precise reference trajectories in outdoor SLAM is GNSS, which was used in the form of integrating the RTK-GPS in the sensor test bed on the Melex vehicle. In order to assess the accuracy of a trajectory estimated to be either visual or LiDAR SLAM the available ground truth trajectory needs to be compared to the estimated one using a proper method. Such methods have been proposed in Ref. [28] and now are widely used in robotics localization research. These are methods for evaluating absolute trajectory error (ATE) and relative pose error (RPE) of sensor positions. The ATE value is the Euclidean distance between the corresponding points of the estimated trajectory and the ground truth trajectory. Thus, the ATE metric allows us to determine how far the estimated position is from the reference position at a certain moment in time on the estimated trajectory. The ATE metric is calculated for the entire trajectory as root mean squared (RMS) error, which allows convenient comparison of results for different SLAM algorithms for the same test data set. Assuming that two trajectories are given: a ground truth and an estimated one with the same number of positions n, and the positions of the sensor are given as homogeneous matrices, the ATE metric for the i-th frame on the trajectory can be determined as:

$$\mathbf{E}_i^{\mathrm{ATE}} = \left(\mathbf{T}_i^{\mathrm{gt}}\right)^{-1}\mathbf{T}_i, \tag{3}$$

and then the ATE value for the entire trajectory is obtained by calculating the RMS error value for all frames. Note that in order to obtain a correct ATE, the trajectories must be converted to a common global coordinate system before the calculation is performed. The need for geometric matching occurs in the case of a trajectory estimated by visual odometry or a monocular visual SLAM algorithm because for algorithms of this kind the trajectory scale s is an unobservable variable [19]. For this reason, the estimated trajectory may exactly reproduce the shape of the reference trajectory, but have a different scale. It is possible to address the scale problem in monocular systems if data from additional sensors are available, in particular inertial measurements from an IMU or AHRS [29]. However, this possibility was not exploited in this project because the stream of images from the smartphone could not be synchronized accurately in time with the AHRS measurements, while the aim was also to compare the trajectories obtained using visual SLAM from the two different types of cameras mounted on the Melex. On the other hand, the correct scale can be easily retrieved by processing the trajectory offline and using the Umeyama algorithm [30] with only several GNSS global pose measurements as reference points.

In the case of LiDAR SLAM, the scale drift problem does not occur. Trajectory matching is performed offline by finding the transformation that minimizes the distance between two groups of rigidly connected points representing these trajectories (vehicle poses), the most commonly used method to accomplish this task is the Umeyama algorithm [30].

On the other hand, the RPE metric determines the difference in transformation (separately the rotational and translational parts) that would exist after following the estimated trajectory and the reference trajectory independently for a given number of frames/images and then calculating the roto-translation between the estimated trajectory and its ground truth trajectory counterpart. The RPE metric for the i-th image frame is given by the formula:

$$\mathbf{E}_i^{\mathrm{RPE}} = \left(\left(\mathbf{T}_i^{\mathrm{gt}}\right)^{-1}\mathbf{T}_{i+1}^{\mathrm{gt}}\right)^{-1}\left(\mathbf{T}_i^{-1}\mathbf{T}_{i+1}\right). \tag{4}$$

Determining the translational or rotational part, one gets the local translational or rotational error measure. The values of the RPE metrics for the entire trajectory are calculated from the corresponding part (translational or rotational) as the RMS error value for all nodes.

The main measure of sensor position estimation error used in the process of evaluating the accuracy of the tested algorithms will be ATE for the i-th frame. The aggregated measure of ATE (RMS) is of lesser importance in this context, since it is not assumed that the resulting maps will be globally accurate, but that the user can get reasonably small errors, for example, translational error smaller than 1 meter, in the coordinate system associated with the local map. This ensures correct recognition of the road structure by the vehicle, which in turn makes it possible to keep on the proper lane and to localize some objects in the vicinity of the road.

In this context, the translational part of the RPE metric (RPE_t) is also useful, as, from the point of view of evaluating the results of the experiments performed, the main advantage of the RPE metric is that it can be directly applied to the evaluation of the trajectory estimated from a monocular camera, in the presented case both the Basler camera and the smartphone's camera. When using the ATE metric, such a possibility does not exist, because, for monocular SLAM, the estimated trajectory and the reference trajectory differ in scale.

Roll and pitch angles of the sensor are of lesser importance, as it is assumed that the vehicle moves on generally flat ground in the urban environment. However, the roll and pitch angles for the reference trajectory can be read from the AHRS relative to the gravity vector. This is used to ensure that the assumption of running on a reasonably flat surface is met for the collected trajectories.

5. Experimental results

Three fully documented experiments were conducted using a pre-calibrated set of sensors on an electric car. The vehicle traveled at speeds ranging from 10 to 30 km/h along various routes in the area of the Warta campus of Poznan University of Technology and the streets adjacent to this area. The routes were documented by plotting ground truth trajectories (obtained from RTK-GPS) on maps from the OpenStreetMaps service (**Figure 5**).

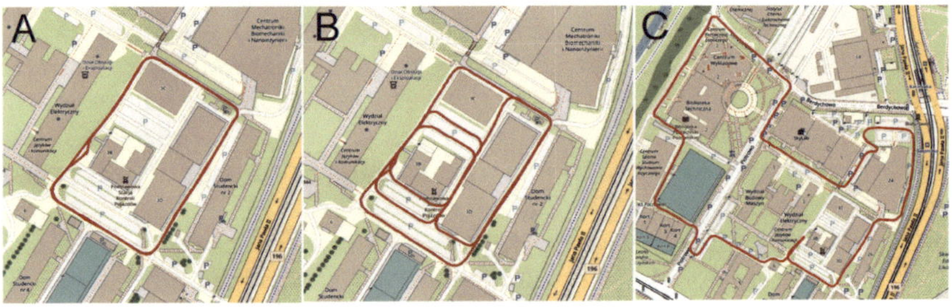

Figure 5.
Ground truth trajectories of three experiments obtained from RTK-GPS and plotted on OpenStreetMaps: Exp. 1 – A single loop inside the campus (A), exp. 2 – Multiple loops inside the campus (B), and exp. 3 – A large loop partially outside the campus (C).

Exp. No.	**1**		**2**		**3**	
RPE$_t$ [m]	**max**	**RMS**	**max**	**RMS**	**max**	**RMS**
LOAM Ouster	1.83	0.53	1.93	0.45	1.14	0.51
ORB-SLAM3 Basler	1.71	0.38	0.92	0.27	0.40	0.18
ORB-SLAM3 Galaxy	0.49	0.23	0.43	0.29	0.42	0.18

Table 1.
Relative translational errors of the trajectory estimated by LOAM (LiDAR SLAM) and ORB-SLAM3 (visual SLAM) for all experiments.

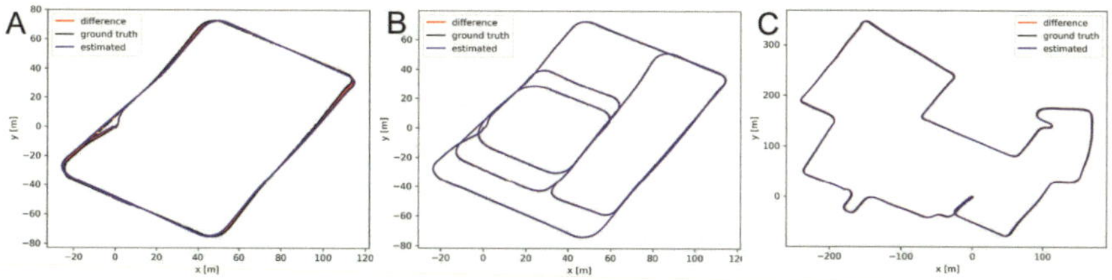

Figure 6.
Plots of the ground truth and estimated trajectories for LOAM/ouster with visualization of the ATE values (red lines) for exp. 1 (A), exp. 2 (B), and exp. 3 (C).

In each run, an estimated vehicle trajectory was obtained from the OS1-128 laser scanner data, the relative position error of which did not exceed 1 m with respect to the RTK-GPS data (**Table 1**). At the same time, all the estimated trajectories are topologically consistent and not distorted (**Figure 6**), also characterized by small absolute ATE errors (the red error lines are faint).

At the same time, the presented experiments confirmed the veracity of the main these—it is possible to obtain the localization of a vehicle equipped with a single passive camera with a relative accuracy of no worse than 1 m (**Table 1**) under conditions similar to the real operating conditions of urban navigation. The difference between the conditions of the experiment and the real conditions came down to the use of a vehicle moving slower than a typical car and the absence of heavy traffic of other vehicles (with the presence of single vehicles and pedestrians).

There are clearly significant differences in the accuracy of the trajectory estimated by ORB-SLAM3 depending on the scenario of a given experiment **Figure** 7. The decisive influence on the quality of localization is the number of point features observed and tracked over a sufficient number of frames. The absence of such features, for example, because of driving at a great distance from objects with distinct

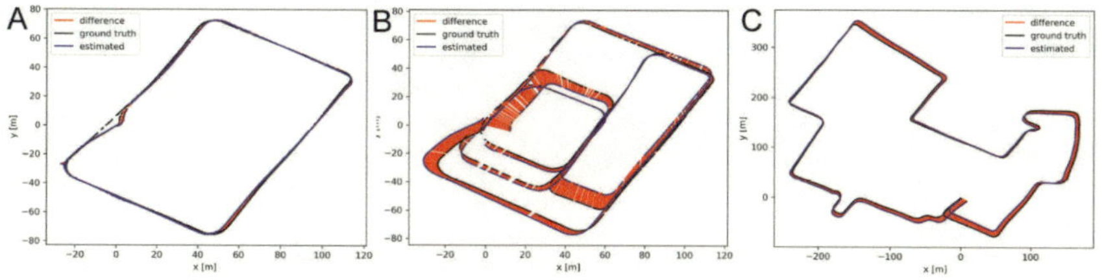

Figure 7.
Plots of the ground truth and estimated trajectories for ORB-SLAM3/Basler with visualization of the ATE values (red lines) for exp. 1 (A), exp. 2 (B), and exp. 3 (C).

surface texture and/or the presence of unstable (weakly repeatable) features generated on trees, causes a sharp increase in local (relative) errors and, consequently, distortion of the estimated trajectory. In the case of the second experiment, in which the vehicle made several passes over the same terrain (loop), it can also be seen that the data from a single camera (monocular visual SLAM) does not allow in some cases to accurately close the loop and completely reduce the drift of the trajectory. In the consequence, the ATE values are much bigger for the vision-based ORB-SLAM3 than for the LiDAR-based LOAM, as shown in **Table 2**.

The differences in the accuracy of the trajectory estimated by ORB-SLAM3 from the low-cost camera of the smartphone measured by the relative measure (**Table 1**) are small, and the obtained values of relative translation errors (both RMS and maximum values) do not differ significantly from analogous values obtained when estimating the trajectory based on images from a professional camera (**Figure 8**).

In the case of experiments 2 and 3 for the Basler camera, the maximum value of the error estimated from the smartphone images is even significantly smaller (0.49 m vs. 1.71 m for the professional camera). This allows us to conclude that the smartphone camera under proper calibration is a sufficient sensor for feature-based monocular SLAM, despite having a rolling shutter and simplified lens. It should be noted that an advantage of the Galaxy A20 smartphone camera (and many others) over the Basler camera with an f = 4 mm lens is a wider angular field of view in the horizontal plane. This can increase the number of matching point features during more rapid changes in vehicle orientation. At the same time, however, the depth of field of the smartphone camera is noticeably smaller than for the classic camera configuration compared here, which in turn leads to the detection of fewer point features overall, especially at greater distances from the vehicle. The number of point features observed and tracked by a sufficient number of frames has a decisive impact on the quality of pose tracking (local localization). Lack of a sufficient number of features results in a sharp increase in local (relative) errors and, consequently, distortion of the estimated trajectory.

Exp. no.	**1**		**2**		**3**	
ATE [m]	**max**	**RMS**	**max**	**RMS**	**max**	**RMS**
LOAM Ouster	2.49	1.26	3.11	2.08	6.33	2.55
ORB-SLAM3 Basler	2.59	1.15	12.90	5.72	12.59	6.53
ORB-SLAM3 Galaxy	7.50	3.04	8.29	4.61	70.83	30.07

Table 2.
Absolute errors of the trajectory estimated by LOAM (LiDAR SLAM) and ORB-SLAM3 (visual SLAM) for all experiments.

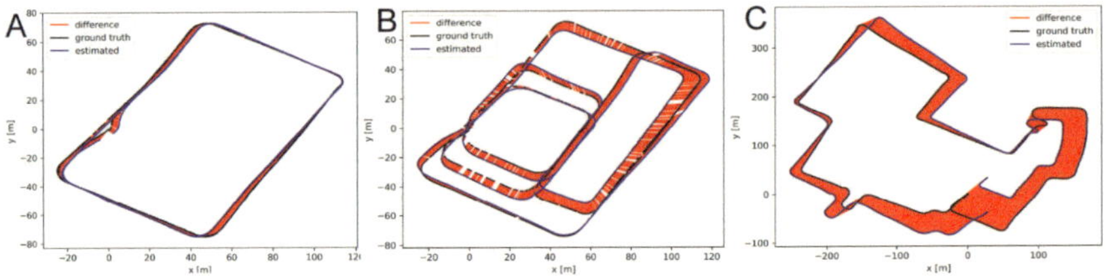

Figure 8.
Plots of the ground truth and estimated trajectories for ORB-SLAM3/galaxy with visualization of the ATE values (red lines) for exp. 1 (A), exp. 2 (B), and exp. 3 (C).

In the case of experiments 2 and 3, during which the vehicle made several loops returning to the same areas, it can also be seen that the data from the Galaxy A20 smartphone camera does not allow, in some cases, for accurate loop closure and trajectory drift reduction. This is particularly evident in **Figure 8C**—ineffective loop closing, probably due to too few features, led to a very large ATE error for the long trajectory (15,288 images were processed), despite the small relative errors.

6. Conclusions

The presented results allow us to conclude that the use of laser scanning to obtain reference trajectories regardless of the availability of the GNSS signal (which often disappears in urban environments) is effective, and it can be applied as part of the methodology for testing vision-based localization systems, as the accuracy obtained is sufficient.

This research has shown that the trajectories estimated by a state-of-the-art visual SLAM system are of acceptable accuracy in the context of local navigation in the urban environment. The obtained accuracy of the vehicle pose estimates is also sufficient for localization of selected objects on the roadsides—such as billboards considered in the CityBrands project.

The multi-loop paths of the presented experiments, which are not present in the publicly available datasets for SLAM, for example, KITTI [25] made it possible to demonstrate the crucial role of topological loop closing in obtaining globally consistent trajectories. A clear decrease of this global consistency was observed for the trajectories obtained from the images acquired using smartphone with its rolling-shutter camera. This can be attributed to the smaller number of point features detected in the lower-quality images that in some cases were insufficient to detect a loop using the bag-of-visual-words approach. This was observed for trajectories that otherwise had local errors not greater than their counterparts estimated from images collected by a professional global-shutter camera. However, supplementing the SLAM algorithm with only locally available (at selected points) GNSS data (which here served only as ground truth) should allow to removal of the observed shortcomings. In general, this research demonstrated that even low-cost commodity hardware can be used to obtain useful trajectories of a vehicle in urban environment if recent algorithms are applied for data processing, and the entire processing pipeline is implemented carefully, focusing on proper calibration of the sensors.

Acknowledgements

This work was funded by POIR.01.01.01-00-0494/20 (CityBrands) grant from the National Centre for Research and Development. The author would like to thank K. Ćwian, A. Banaszczyk, and K. Żywanowski for their help in conducting the presented experiments.

Author details

Piotr Skrzypczyński
Institute of Robotics and Machine Intelligence, Poznań University of Technology, Poznań, Poland

*Address all correspondence to: piotr.skrzypczynski@put.poznan.pl

References

[1] Cadena C, Carlone L, Carrillo H, Latif Y, Scaramuzza D, Neira J, et al. Past, present, and future of simultaneous localization and mapping: Toward the robust-perception age. IEEE Transactions on Robotics. 2016;**32**(6): 1309-1332

[2] Scaramuzza D, Fraundorfer F. Visual odometry: Part I the first 30 years and fundamentals. IEEE Robotics & Automation Magazine. 2011;**18**(4):80-92

[3] Skrzypczyński P. LiDAR localization and mapping for autonomous vehicles: Recent solutions and trends. In: Automation 2021: Recent Achievements in Automation, Robotics and Measurement Techniques, AISC 1390. Cham: Springer; 2021. pp. 251-261

[4] Campos C, Elvira R, Gómez Rodríguez JJ, Montiel JMM, Tardós JD. ORB-SLAM3: An accurate open-source library for visual, visual-inertial and multi-map SLAM. IEEE Transactions on Robotics. 2021;**37**(6):1874-1890

[5] Zhang J, Singh S. Low-drift and real-time LiDAR odometry and mapping. Autonomous Robots. 2017;**41**(2): 401-416

[6] Quigley M, Conley K, Gerkey B, Faust J, Foote T, Leibs J et al. ROS: An Open-Source Robot Operating System, Workshops at the IEEE International Conference on Robotics and Automation. Kobe: IEEE; 2009

[7] Rehder J, Siegwart R, Furgale P. A general approach to spatiotemporal calibration in multisensor systems. IEEE Transactions on Robotics. 2016;**32**(2): 383-398

[8] Nowicki MR. Spatiotemporal calibration of camera and 3D laser scanner. IEEE Robotics and Automation Letters. 2020;**5**(4):6451-6458

[9] Zhang Q, Pless R. Extrinsic calibration of a camera and laser range finder (improves camera calibration). In: IEEE/RSJ International Conference on Intelligent Robots and Systems (IROS), Sendai. IEEE; 2004. pp. 2301-2306

[10] Takasu T. RTKLIB: Open Source Program Package for RTK-GPS, FOSS4G 2009. Tokyo: Free and Open Source Software for Geospatial (FOSS4G), Open Source Geospatial Foundation; 2009

[11] Durrant-Whyte H, Bailey T. Simultaneous localization and mapping: Part I. IEEE Robotics and Automation Magazine. 2006;**13**(2):99-110

[12] Strasdat H, Montiel JMM, Davison AJ. Visual SLAM: Why filter? Image and Vision Computing. 2012; **30**(2):65-77

[13] Grisetti G, Kümmerle R, Stachniss C, Burgard W. A tutorial on graph-based SLAM. IEEE Intelligent Transportation Systems Magazine. 2010;**2**(4):31-43

[14] Kschischang F, Frey B, Loeliger H-A. Factor graphs and the sum-product algorithm. IEEE Transactions on Information Theory. 2001;**47**(2):498-519

[15] Nowicki M, Skrzypczyński P. Indoor navigation with a smartphone fusing inertial and WiFi data via factor graph optimization. In: Mobile Computing, Applications, and Services. MobiCASE 2015, LNCS 162. Cham: Springer; 2015. pp. 280-298

[16] Mur-Artal R, Montiel JMM, Tardós JD. ORB-SLAM: A versatile and

accurate monocular SLAM system. IEEE Transactions on Robotics. 2015;**31**(5): 1147-1163

[17] Ćwian K, Nowicki M, Wietrzykowski J, Skrzypczyński P. Large-scale LiDAR SLAM with factor graph optimization on high-level geometric features. Sensors. 2021;**21**(10):3445

[18] Kümmerle R, Grisetti G, Strasdat H, Konolige K, Burgard W. G^2o: A General Framework for Graph Optimization. Proc. IEEE Int. Conf. on Robotics & Automation: Shanghai; 2011. pp. 3607-3613

[19] Strasdat H. Local Accuracy and Global Consistency for Efficient Visual SLAM [PhD Dissertation]. London: Imperial College; 2012

[20] Triggs B, McLauchlan PF, Hartley RI, Fitzgibbon AW. Bundle Adjustment – A Modern Synthesis, Vision Algorithms: Theory and Practice, LNCS 1883. Berlin, Heidelberg: Springer; 2000. pp. 298-372

[21] Konolige K. Sparse sparse bundle adjustment. In: Proc. British Machine Vision Conference. Aberystwyth: British Machine Vision Association; 2010. pp. 102.1-102.11

[22] Segal A, Haehnel D, Thrun S. Generalized-ICP. In: Proc. Robotics: Science and Systems. Seattle: RSS Foundation; 2009

[23] Schmidt A, Kraft M, Fularz M, Domagala Z. Comparative assessment of point feature detectors and descriptors in the context of robot navigation. Journal of Automation, Mobile Robotics & Intelligent Systems. 2013; 7(1):11-20

[24] Galvez-López D, Tardos JD. Bags of binary words for fast place recognition in image sequences. IEEE Transactions on Robotics. 2012;**28**(5):1188-1197

[25] Geiger A, Lenz P, Urtasun R. Are we ready for autonomous driving? The KITTI vision benchmark suite. In: IEEE Conference on Computer Vision and Pattern Recognition. Providence: IEEE; 2012. pp. 3354-3361

[26] Rublee E, Rabaud V, Konolige K, Bradski G. ORB: An Efficient Alternative to SIFT or SURF. Barcelona: IEEE Int. Conf. on Computer Vision; 2011. pp. 2564-2571

[27] Będkowski J. Large-Scale Simultaneous Localization and Mapping. Springer, Singapore: Cognitive Intelligence and Robotics; 2022

[28] Sturm J, Engelhard N, Endres F, Burgard W, Cremers D. A benchmark for the evaluation of RGBD SLAM systems. In: IEEE/RSJ International Conference on Intelligent Robots and Systems (IROS), Vilamoura: IEEE; 2012. pp. 573-580

[29] Nützi G, Weiss S, Scaramuzza D, Siegwart R. Fusion of IMU and vision for absolute scale estimation in monocular SLAM. Journal of Intelligent and Robotic Systems. 2011;**61**:287-299

[30] Umeyama S. Least-squares estimation of transformation parameters between two point patterns. IEEE Transactions on Pattern Analysis and Machine Intelligence. 1991;**13**(4): 376-380